SCIENCE, ORDER, AND CREATIVITY

BANTAM NEW AGE BOOKS

This important imprint includes books in a variety of fields and disciplines and deals with the search for meaning, growth and change. They are books that circumscribe our times and our future.

Ask your bookseller for the books you have missed.

THE MIND'S I by Douglas R. Hofstadter and Daniel C. Dennett
NATURAL ESP: THE ESP CORE AND ITS RAW
   CHARACTERISTICS by Ingo Swann
THE NEW STORY OF SCIENCE by Robert M. Augros and
   George N. Stanciu
ON HUMAN NATURE by Edward O. Wilson
ORDER OUT OF CHAOS by Ilya Prigogine and Isabelle Stengers
ORIGINS: A SKEPTIC'S GUIDE TO THE CREATION OF LIFE ON
   EARTH by Robert Shapiro
PERFECT SYMMETRY by Heinz R. Pagels
PROSPERING WOMAN by Ruth Ross
SCIENCE, ORDER, AND CREATIVITY by David Bohm and
   F. David Peat
SHAMBHALA: THE SACRED PATH OF THE WARRIOR by
   Chogyam Trungpa
SPACE-TIME AND BEYOND (The New Edition) by Bob Toben and
   Fred Alan Wolf
STAYING SUPPLE by John Jerome
SUPERMIND by Barbara B. Brown
SYMPATHETIC VIBRATIONS: REFLECTIONS ON PHYSICS AS A
   WAY OF LIFE by K. C. Cole
SYNCHRONICITY: THE BRIDGE BETWEEN MATTER AND MIND
   by F. David Peat
THE TAO OF LEADERSHIP by John Heider
THE TAO OF PHYSICS, (Revised Edition) by Fritjof Capra
TO HAVE OR TO BE? by Erich Fromm
THE TURNING POINT by Fritjof Capra
THE WAY OF THE SHAMAN: A GUIDE TO POWER AND
   HEALING by Michael Harner
ZEN AND THE ART OF MOTORCYCLE MAINTENANCE by
   Robert M. Pirsig

SCIENCE, ORDER, AND CREATIVITY

DAVID BOHM
F. DAVID PEAT

BANTAM BOOKS
NEW YORK · TORONTO · LONDON · SYDNEY · AUCKLAND

SCIENCE, ORDER, AND CREATIVITY
A Bantam Book / November 1987

Line art #10–15 (pages 122, 143, 152 and 153) reproduced (by permission) by Benoit Mandelbrot, from THE FRACTAL GEOMETRY OF NATURE, W. H. Freeman, N.Y., 1982. Copyright by B. Mandelbrot.

New Age and the accompanying figure design as well as the statement "the search for meaning, growth and change" are trademarks of Bantam Books.

Book Design by Ann Gold.

Library of Congress Cataloging-in-Publication Data

Bohm, David.
 Science, order, and creativity.

 Includes bibliographical references and index.
 1. Science—Philosophy. 2. Creative ability in science. 3. Order (Philosophy) I. Peat, F. David, 1938– II. Title.
 Q175.B666 1987 501 87-11448
 ISBN 0-553-34449-8 (pbk.)

Published simultaneously in the United States and Canada

Bantam Books are published by Bantam Books, a division of Bantam Doubleday Dell Publishing Group, Inc. Its trademark, consisting of the words "Bantam Books" and the portrayal of a rooster, is Registered in U.S. Patent and Trademark Office and in other countries. Marca Registrada. Bantam Books, 666 Fifth Avenue, New York, New York 10103.

CONTENTS

INTRODUCTION

*T*his book grew out of a series of dialogues that took place between us over the last fifteen years. It therefore seems appropriate, in this Introduction, that the reader should be given some idea of the genesis of our book and of the kinds of thoughts and questions that stirred us into writing it. Since this naturally involves our own personal backgrounds, feelings, and attitudes, it is most easily presented in the form of a dialogue between us. Indeed, what follows could well have taken place during one of our afternoon walks together while the book was being written.

DAVID BOHM: I think that it would be a good idea to begin with the book itself. What first led you to suggest that we should write a book together?

DAVID PEAT: Well, a question like that takes me right back to my childhood. You see, as far back as I can remember, I was always interested in the universe. I can still remember standing under a street lamp one evening—I must have been eight or nine—and looking up into the sky and wondering if the light went on forever and ever, and what it meant for something to go on forever and ever, and if the universe ever came to an end. You know the sorts of questions. Well, pretty soon the idea began to excite me that the human mind was able to ask these sorts of questions and in some way comprehend the vastness of everything.

These sorts of ideas continued right through school, along with a feeling of the interconnectedness of everything. It was almost as if the entire universe were a living entity. But of course, when I got down to the serious business of studying science at university, all this changed. I felt that the deepest questions, particularly about the quantum theory, were never properly answered. It seemed pretty clear that most scientists were not really interested in these sorts of questions. They felt that they were not really related to their day-to-day research. Instead, we were all encouraged to focus on getting concrete results that could be used in published papers and to work on problems that were "scientifically acceptable." So fairly early on, I found myself getting into hot water because I was always more excited by questions that I didn't know how to answer than by more routine research. And of course, that's not the way to build up an impressive list of scientific publications.

DAVID BOHM: But you were not simply interested in science alone?

DAVID PEAT: No, I was attracted to music, theater, and the visual arts. I could see that they were another very important way of responding to nature and understanding our position in the universe. I always felt that, in some deeper sense, the really important figures in science and the arts were fundamentally doing the same thing and responding to the same ultimate origin. This essential relationship between the sciences and the arts is still very important to me.

But with the exception of a few good friends, it was difficult to find people who shared my enthusiasms. I had begun a kind of indirect dialogue with you by reading your papers and I sensed a similar interest. The end result was that in 1971 I took a year's sabbatical leave to come to Birkbeck College in London so that we could explore all these things together.

DAVID BOHM: Yes, I remember we met once or twice a week and talked well into the evening.

DAVID PEAT: Do you remember how I began by asking you scientific questions but very soon we moved into the whole area of consciousness, society, religion, and culture? After I returned to Canada, it was clear that we should go on meeting again on a fairly regular basis to continue our dialogues.

DAVID BOHM: Yes, but it also began to emerge that the dialogue itself was the key issue. And that this was intimately related to all the others. The essential question was; How can we engage in dialogue in a creative way?

DAVID PEAT: Yes, and I think this was what eventually led me to suggest that we should write a book together. In a sense, working on this book has become a continuation of our dialogue. Of course, many of the ideas we're going into really began with you.

DAVID BOHM: Yes, but in exploring them through dialogue they began to develop in new ways and it also became possible to communicate them more clearly.

DAVID PEAT: Communication plays a very important role in how new ideas can develop. In fact, this whole project has been a very exciting one.

DAVID BOHM: I think this has come out of the intense interest we both have in going into these questions. You see, I, too, felt that kind of wonderment and awe in my early days, along with an intense wish to understand everything, not only in detail but also in its wholeness.

I learned later that many of my fundamental interests were what other people called philosophical and that scientists tended to look down on philosophy as not being very serious. This created a problem for me, as I was never able to see any inherent separation between science and philosophy. Indeed in earlier times, science was called *natural philosophy* and this corresponded perfectly with the way I saw the whole field. At university, I had a few friends who approached the subject in the same way and we had many discussions in a spirit of friendship and common inquiry. However, in graduate school at the California Institute of Technology, which I entered in 1939, I found that there was a tremendous emphasis on competition and that this interfered with such free discussions. There was a great deal of pressure to concentrate on learning formal techniques for getting results. It seemed that there was little room for the desire to understand in the broad sense that I had in mind. Neither was there a free exchange and the friendship that is essential for such understanding.

Although I was quite capable of mastering these mathematical techniques, I did not feel that it was worth going on with, not without a deeper philosophical ground and the spirit of common inquiry. You see, it is these very things that provide the interest and motivation for using mathematical techniques to study the nature of reality.

DAVID PEAT: But things did improve for you when you went to Berkeley, didn't they?

DAVID BOHM: Yes, when I went to work with J. Robert Oppenheimer, I found a more congenial spirit in his group. For example, I was introduced to the work of Niels Bohr and this stimulated my interest, especially in the whole question of the oneness of the observer and the observed. Bohr saw this in the context of the undivided wholeness of the entire universe. I can still recall the many discussions I had on matters like this which had the effect of setting me on the course I'm still following today. Philosophy played an inseparable part, but it was not just a matter of philosophizing about material that was already present in science in a more or less finished form. I was more interested in broader questions which have been the very source and origin not only of my interest, but also of many key ideas which later took mathematical form.

DAVID PEAT: Maybe you could give an example of this relationship of scientific ideas to the underlying philosophy.

DAVID BOHM: When I worked at the Lawrence Laboratory, after taking my Ph.D., I became very interested in the electron plasma. This is a dense gas of electrons that exhibits radically different behavior from the other, normal states of matter and it was a key to much of the work the laboratory was doing at the time. My insights sprang from the perception that the plasma is a highly organized system which behaves as a whole. Indeed in some respects, it's almost like a living being. I was fascinated with the question of how such organized collective behavior could go along with the almost complete freedom of movement of the individual electrons. I saw in this an analogy to what society could be, and perhaps as to how living beings are organized. Later, when I went to Princeton, I extended this view in order to treat electrons in metals on the same footing.

DAVID PEAT: But I think that you were also a little disturbed at the way your results were being used.

DAVID BOHM: Well, I had worked out a number of equations and formulae and some of these played a key part in research into fusion and into the theory of metals. But a few years later, when I attended some scientific conferences, it became clear that these formulae had been taken up and transformed into more abstract forms, while the ideas behind them were ignored. People didn't even seem to want to talk about the ideas. The general spirit was that the main aim of physics is to produce formulae that will correctly predict the results of experiments. In the face of all this, I began to feel that there was no point in going on with the plasma research and so I lost interest in it.

However, I did continue to work in physics and developed the causal interpretation of the quantum theory and the implicate order. But both of these originated, to a large extent, in philosophical questions.

DAVID PEAT: As a matter of fact, these early papers of yours were just what first excited my interest. I started my first research by looking at systems of many electrons, and I was soon interested in the relationship between individual and collective behavior. Of course, it was your papers that helped me to obtain some insights into the relationship between the individual and the whole. I think they also gave me the confidence to go further and try to look a little deeper into questions about the foundation of quantum theory. But as I said before, the overwhelming climate was unsympathetic to such an approach. I could see that most physicists could not see the point you were driving at.

DAVID BOHM: They seemed more interested in the formulae than the ideas behind them.

DAVID PEAT: But this leads me to what I think is a very important question. What would you say to the prevailing belief that the mathematical formalism itself expresses the very essence of our knowledge of nature?

DAVID BOHM: Of course, some scientists, notably the Pythagoreans, held views like this in ancient times. And others, like Kepler, believed that mathematics was a basic source of truth. But this notion that the mathematical formalism expresses the essence of our knowledge about nature did not really become commonly accepted until relatively recent times. For example, when I was a student, most physicists felt that a physical or intuitive concept was the essential point and that the mathematical formalism had to be understood in relation to this.

DAVID PEAT: But how did this emphasis on mathematics come about?

DAVID BOHM: It was really because the quantum theory, and to a lesser extent relativity, were never understood adequately in terms of physical concepts that physics gradually slipped into the practice of talking mostly about the equations. Of course, this was because equations were the one part of the theory that everyone felt they could really understand. But this inevitably developed into the notion that the equations themselves are the essential content of physics. To some extent this began as early as the 1920s when the astronomer Sir James Jeans proposed that God must be a mathematician.[1] Heisenberg later gave it an enormous boost with his idea that science could no longer visualize atomic reality in terms of physical concepts and that

1. Sir James Jeans, *The Mysterious Universe* (New York: Cambridge University Press, 1930).

mathematics is the basic expression of our knowledge of reality.[2] Along with this went a whole change in the notion of what was meant by an intuitive or imaginative grasp. This had previously been identified with the ability to visualize ideas and concepts, but now Heisenberg was claiming that intuition and imagination provide not a picture of reality but a mental display of the meaning of the mathematics.

Now I don't agree with these developments. In fact, I feel that the current emphasis on mathematics has gone too far.

DAVID PEAT: But on the other hand, many of the deepest scientific thinkers have used criteria of mathematical beauty in the development of their theories. They believed that the deepest scientific explanations must also be mathematically beautiful. Without the requirement of mathematical aesthetics a great many discoveries would not have been made. Surely in your own work the criteria of mathematical elegance must have acted as a signpost that you were on the right track?

DAVID BOHM: Certainly mathematics gives rise to creative insights, and the search for mathematical beauty can be a helpful guide. Scientists who have worked in this way have often been successful in deriving new knowledge through an emphasis on the mathematical formalism. I have already mentioned Kepler and Heisenberg, and in modern times I could add such names as Dirac, von Neumann, Jordan, and Wigner. But mathematics was never the *sole* criterion in their discoveries. Moreover, this does not mean that *everyone* thinks alike in this regard. In fact, I think that verbal concepts, pictorial aspects, and philosophical thinking can contribute significantly to new ideas. Einstein certainly appreciated mathematical beauty very keenly but he did not actually begin from the mathematics, especially in his most creative period. Instead, he started with unspecifiable feelings and a succession of images out of which more detailed concepts eventually emerged. I would go along with this and add that ideas arising in this way, or in other ways, may

2. A. Miller, *Imagery in Scientific Thought*, Birkhauser, Boston, Bern, and Stuttgart, 1984.

eventually lead to further mathematical developments and even to the suggestion of new forms of mathematics.

It seems arbitrary to say that mathematics must play a unique role in the expression of reality. Mathematics is only one function of the human mind, and other functions can surely be just as important—even in physics.

DAVID PEAT: This dialogue is moving in an interesting direction. We seem to be saying that physics may have taken a wrong direction in giving so much emphasis to its formalisms. But I'm sure that many scientists would point out that mathematics happens to be the most abstract and logically coherent way of thinking that is known to us. It seems to be totally open to free creation and not bounded by the requirements of sense experiences of ordinary reality. Doesn't that give it a unique status?

DAVID BOHM: Well, in reply, I'd like to bring in the work of Alfred Korzybski, an American philosopher who was fairly well known in the early twentieth century.[3] He said that mathematics is a limited linguistic scheme, which makes possible great precision and coherence—but at the expense of such extreme abstraction that its applicability has, in certain key ways, to be bounded.

Korzybski said, for example, that whatever we say a thing is, it isn't. First of all, whatever we say is words, and what we want to talk about is generally not words. Second, whatever we *mean* by what we say is not what the thing actually is, though it may be similar. For the thing is always *more* than what we mean and is never exhausted by our concepts. And the thing is also *different* from what we mean, if only because no thought can be absolutely correct when it is extended indefinitely. The fact that a thing has qualities going beyond whatever we think and say about it is behind our notion of objective reality. Clearly, if reality were ever to cease to show new aspects that are not in our thought, then we could hardly say that it had an objective existence independent of us.

---

3. A. Korzybski, *Science and Sanity*, International Neo-Aristotelian Publishing Company, Lakeville, Conn., 1950.

All this implies that *every* kind of thought, mathematics included, is an abstraction, which does not and cannot cover the whole of reality. Different kinds of thought and different kinds of abstraction may together give a better reflection of reality. Each is limited in its own way, but together they extend our grasp of reality further than is possible with one way alone.

DAVID PEAT: What you have said about Korzybski reminds me of René Magritte's painting of a pipe which also contains the words *This is not a pipe*. However realistic a painting may be, it falls indefinitely short of being an actual pipe. And ironically, the word *pipe* in the title is not an actual pipe either. Perhaps, in the spirit of Magritte, every theory of the universe should have in it the fundamental statement "This is not a universe."

DAVID BOHM: Actually, a theory is a kind of map of the universe, and like any other map, it is a limited abstraction and not entirely accurate. Mathematics provides one aspect of the overall map, but other ways of thinking are needed along the lines we have been discussing.

DAVID PEAT: Well, it's certainly true that in the early days of quantum theory, the leading physicists like Bohr, Heisenberg, Pauli, Schrodinger, and de Broglie were vitally concerned with philosophical questions, especially on the relationship between ideas and reality.

DAVID BOHM: These questions go beyond the limited scope of physics as it is generally known today. Each of these thinkers approaches the problem in his own way and there are important and subtle differences between them that we tend to overlook today. But the general practice of physics has indeed become remote from these deeper considerations. It tends to concentrate on technical questions, and for this reason, it seems to have lost contact with its own roots. For example, in any attempt to unify quantum mechanics and relativity, especially general relativity, there are fundamental questions that must be faced. How can physicists hope to work successfully in this field when they ignore the subtle and unresolved problems that still lie buried in the early period of quantum mechanics?

DAVID PEAT: I remember that these sorts of questions kept coming up in the discussions we had together at Birkbeck College. We were especially concerned with the narrowness of vision that is developing, not only in physics, but quite generally in scientific research.

DAVID BOHM: We used an analogy from human vision. The details of what we see are picked up in a small central part of the retina called the fovea. If this is destroyed, then detailed vision is lost, but general vision, which comes from the periphery of the retina, remains. But if the periphery is damaged, while the fovea remains intact, even the details lose all their meaning. By analogy, we asked if science was in danger of suffering a similar "damage" of its vision. By giving so much emphasis on mathematics, science seems to be losing sight of the wider context of its vision.

DAVID PEAT: But originally there *was* such a general vision of the universe, humanity, and our place in the whole. Science, art, and religion were never really separate.

DAVID BOHM: But as time went on, this vision changed with specialization. It grew progressively narrower and eventually led to our present approach, which is, in large part, fragmentary. I think this development partly arose because physics had become the pattern or ideal toward which all the sciences aim. While most sciences are not as dominated by mathematics, the essential point is the spirit with which mathematics tends to be done. Its general aim is to try to analyze everything into independent elements that can be dealt with separately. This encourages the hope that any problem can be split off into a separate fragment. Now it is still true that science also contains a movement toward synthesis and to discovering broader contexts and more general laws. But the prevailing attitude has been to put the major emphasis on analysis and on splitting off the key factors of each situation. Scientists hope that this will enable them to extend their powers indefinitely to predict and control things.

DAVID PEAT: It's important to emphasize that not only is this sort of approach fundamental to physics but it also extends into chemistry, biology, the neurosciences, and even into economics and psychology.

DAVID BOHM: By concentrating on this sort of analysis and constantly splitting off problems into specialized areas, we increasingly ignore the wider context that gives things their unity. In fact, this spirit is now spreading beyond science, not only into technology, but into our general approach to life as a whole. Understanding is now valued as the means to predict, control, and manipulate things. Of course, beginning with Francis Bacon, this has always been important but never so dominant as today.

DAVID PEAT: Yes, science has been moving at an ever-increasing rate since the nineteenth century and it's bringing with it a host of technological changes. But it is only relatively recently that so many people have begun to question if all this progress is really beneficial. We're beginning to realize that the cost of progress is more and more specialization and fragmentation to the point where the whole activity is losing its meaning. I think that the time has come for science to pause and take a careful look at where it is going.

DAVID BOHM: I think that even more than this we need to change what we mean by "science." The moment has come for a creative surge along new lines. This is essentially what we are proposing in *Science, Order, and Creativity*.

DAVID PEAT: But most scientists would be shocked by such a suggestion. After all, it must look as if science has never been more active and successful than it is today. In every field there are new frontiers opening up and new technologies are being exploited. Think of all those novel experimental techniques, exciting new theories, and interesting problems for an indefinite number of research workers to tackle. Take medicine, for example: So many diseases have been wiped out and there is the promise of eradicating even more. And there are the new fields of biotechnology and genetic engineering, and let's not forget the changes that are being made by computers and mass communication. In every area of life, science is making a really powerful impact.

DAVID BOHM: All this is true, but some very important factors have been neglected in obtaining such progress. First of all, there has been an overall fragmentation in our general attitude to reality. This leads us to focus always on particular problems, even when they are significantly related to a broader context. As a result, we fail to notice the unforeseen negative consequences, which cannot always be dealt with in terms of a fragmentary mode of thought. The result is that these difficulties spread into the whole context and eventually come back to create problems that may be worse than those we started with. For example, by exploring natural resources in a fragmentary manner, society has brought about the destruction of forests and agricultural lands, created deserts, and even threatens the melting of the ice caps.

DAVID PEAT: I remember looking into the whole question of the development of more productive strains of crops. It's not at all clear that they have been totally beneficial. To begin with, it leads to the problem of the great vulnerability of a limited genetic strain, and there is an increased reliance on fertilizers, pesticides, herbicides, and ripeners. When you take all this, together with the more efficient farming techniques that these new crops require, it produces radical changes in agrarian societies that have to rely more and more on an industrial basis. In the end, the whole society changes in an uncontrolled way and its economy becomes dependent on imports and it is vulnerable to global instabilities.

DAVID BOHM: Of course, many people think that solving these sorts of problems is only a matter of studying ecology or some other speciality. Certainly ecology does begin to acknowledge the complex dependence of each activity on the whole context. But really the problem is as much one of economics as it is of ecology, and this leads on to politics, and to the structure of society and the nature of human beings in general.

The key issue is this: How is it possible to subject all these factors to prediction and control in order to manipulate the system and bring about good order? Clearly this is an impossible demand. To begin with, there is the infinite complexity involved, and the extreme instability of these systems, which requires almost perfect and, probably unattainable, degrees of control. But more important, the system itself depends on human beings. And how can science lead human beings to control themselves? How do scientists propose to control hatred between nations, religions, and ideologies when science itself is fundamentally limited and controlled by these very things? And what about the growing psychological tension in a society that is so unresponsive to basic human needs that life seems, for many, to have lost its meaning? In the face of all this, some people break down mentally, or become dependent on various drugs, while others engage in mindless violence.

DAVID PEAT: It seems impossible to dream that through some sort of new discovery in chemistry or biology or the behavioral sciences that these problems will ever be brought under control. They are so far-reaching and pervasive. How does science intend to end the danger of mutual annihilation that exists in the world? After all, this has its origin in the fear, mistrust, and hatred between nations. It seems to me that the more science and technology develop, the more dangerous the whole situation is becoming.

DAVID BOHM: Of course, a century or so ago the benefits from science generally outweighed the negative effects, even when the whole endeavor was carried out without regard for long-range consequences. But the modern world is finite and we have almost unlimited powers of destruction. It's clear that the world has passed a point of no return. This is one reason why we have to pause and consider the possibility of a fundamental and extensive change in what science means to us.

DAVID PEAT: What we need is not so much new scientific ideas, although these are still going to be of great interest. The question is how can science, when it is based on a fragmentary attitude to life, ever understand the essence of real problems that depend on an indefinitely wide context? The answer does not lie in the accumulation of more and more knowledge. What is needed is *wisdom*. It is a lack of wisdom that is causing most of our serious problems rather than a lack of knowledge.

DAVID BOHM: But this also implies goodwill and friendliness. This seems to be lacking today, among scientists as much as in the general public.

DAVID PEAT: Of course, goodwill and friendship are important if people are going to work together for the common good. But in the long run, I think that we may have to sacrifice some of the values that we hold so dear today. For example, we have to question the indefinite increase of individual comfort and prosperity and the preeminence of the competitive spirit, which is basically divisive and fragmentary.

DAVID BOHM: Yes, and it's arbitrary to forever limit science to what it has become today. After all, this was the result of a historical process that involved many fortuitous elements. We have to explore in a creative way what a new notion of science might be, a notion that is suitable for our present time. This means that all the subjects that we have been talking about will have to come into the discussion.

DAVID PEAT: I think that if we are to understand this call for a new creative surge in science, then we also have to understand the historical perspective that you've been talking about. We need to understand how our present fragmentary approach came about. For example, it would be interesting to think about what would have happened if different pathways that were available at the time had been fully explored in the past.

DAVID BOHM: But these sorts of discussion cannot be restricted to science alone. We have to include the whole range of human activities. Our aim is to throw light on the nature of creativity and how it can be fostered, not only in science but in society, and in the life of each individual. This is the ultimate nature of the surge we are calling for.

ONE

REVOLUTIONS, THEORIES, AND CREATIVITY IN SCIENCE

Science today is exerting an ever-increasing influence over the world's societies, yet at its very heart, it is beset with serious difficulties. One of the most pervasive of these involves its fragmentary approach to nature and reality. In the Introduction it was suggested that, in an age in which science is taken to be the key to increasing progress and the betterment of life, this fragmentary approach can never resolve the deeper problems which now face our world. Many of these problems depend on contexts so broad that they ultimately extend into the whole of nature, society, and the life of each individual. Clearly such problems can never be solved within the limited contexts in which they are normally formulated.

It is only by moving beyond its present fragmentation that science can hope to make a realistic contribution to these, more serious problems which face us. But fragmentation should not be confused with the act of division of an area of knowledge into particular fields of specialization or with the abstraction of specific problems for study. These divisions may be perfectly legitimate, and in fact, they are an essential feature of science. Rather, as the term indicates, *to fragment* means "to break up or smash." Fragmentation therefore arises when an attempt is made to impose divisions in an arbitrary fashion, without any regard for a wider context, even to the point of ignoring essential connections to the rest of the world. The image of a watch that has been smashed by a hammer comes to mind, for what results is not an appropriate set of divisions but arbitrary fragments which have little or no significance to the working of the watch. Many of our current attempts to deal with the serious problems discussed in the Introduction result in solutions and actions which are as fragmentary and irrelevant as the parts of a broken watch.

FRAGMENTATION AND CHANGE IN SCIENCE

Science is an attempt to understand the universe and humanity's relationship to nature. How then is it possible for fragmentation to grow out of such an approach? The very notion of scientific understanding appears to be totally incompatible with a fragmentary attitude to reality. To understand how the fragmentary approach to the problems and difficulties discussed in the Introduction came to pervade the whole of science, it is first necessary to understand not only what fragmentation means, but also how it operates in practice. This involves particularly subtle and complex issues. To begin with, it is important once again to distinguish fragmentation from simple specialization and from the practical division of knowledge into various subdivisions. Clearly some such form of specialization was the essential step in the development of civilization.

Indeed, the study of any field begins with a natural act of abstraction, in order to focus on certain features of interest. To be able to give attention to something, it is first necessary to abstract or isolate its main features from all the infinite, fluctuating complexity of its background.

When such an act of perceptive abstraction is free from an excessive mechanical rigidity, then it does not lead to fragmentation, but rather it reflects the ever-changing relationship of the object to its background. In order to recognize a face in a moving crowd, for example, an act of perceptive abstraction is required in which important features are isolated and integrated together. In a similar way, nonrigid focusing of the mind upon a field of interest will allow a corresponding integration of all relevant features in this field.

As a simple illustration, think of an intern who examines a patient in the emergency ward of a hospital. This doctor must make a preliminary diagnosis based on characteristic signs and symptoms that have to be abstracted from the infinite variety of appearances and behavior of the patient. This diagnosis therefore relies upon an essential division and classification of groups of symptoms and findings. But this division must never be fixed rigidly beforehand. Rather, the doctor must constantly check and confirm his or her hypotheses, changing them when they are not confirmed.

The preliminary diagnosis may point to some trauma in a localized region of the body, the dysfunction of an organ, a generalized infection, or some overall disorder of the metabolism. The recognition of a specific disease therefore depends upon the doctor's ability to recognize a whole picture of symptoms which have been abstracted out of a complex background. Given this diagnosis, the intern will then call upon the assistance of a doctor who specializes in one of the subdivisions of medicine, for example injuries to the brain, disorders of the gastrointestinal system, fractures of the bones, infectious diseases. When medicine works well, it combines this generalized knowledge with the more focused and detailed knowledge of the specialist. The danger, however, is always present that by converging upon a particular symptom, or area of the body, its connection with the larger whole of the patient's lifestyle and the lifestyle of the whole society may be neglected. When this happens, the deeper nature of the disorder is obscured and specialization gives way to fragmentation, which will lead to inappropriate treatment.

In a similar way, science has developed into a number of general areas, such as physics, chemistry, and biology. In turn, each of these fields is further broken down into more specific specializations. Physics, for example, includes elementary particles, nuclear, atomic, molecular, condensed matter, fluids, astrophysics, and so on. Each particular discipline involves its own highly specific areas of knowledge together with characteristic theoretical and experimental approaches that have been built up through historical evolution.

In the seventeenth century, for example, the study of gases embraced both physics and chemistry, and a range of different approaches and experimental techniques were used in the one laboratory. The Irishman Robert Boyle, for example, was interested in the behavior of gases, both chemical and physical without distinction. In particular he became fascinated by what he called the "spring" of a gas, the way its volume changes with pressure. In order to make careful measurements of this relationship, it was necessary to isolate each particular gas from background contingencies, such as changes in temperature. But once Boyle's law had been established, it became possible to widen the investigation and to explore the effect of both pressure and temperature on the same volume of gas. In turn, ever more refined experiments could detect deviations on the part of individual gases, from this ideal behavior. But by now, the study of gases had divided itself into two main areas, their physical and their chemical behaviors, which were studied by scientists with quite different backgrounds and trainings.

The example of Boyle's research illustrates two particular tendencies in specialization: first, that a topic of general interest, in this case the behavior of gases, can become divided into several distinct fields of study; and second, the way in which a scientific investigation proceeds by focusing, through carefully designed experiments, upon some particular property of a system and then attempting to study it in isolation from the wider context of its environment. Once this particular property is fully understood, the context can then be expanded to include additional effects and properties. Ideally, areas of specialization are never rigidly fixed but evolve dynamically, in a state of flux, subdividing into narrow regions of specialization at one time then becoming more generalized at another. Provided that these boundaries remain fluid and scientists are aware of the wider context of each experiment and concept, then the problems of fragmentation need not arise.

But, in general, science today is becoming more and more specialized so that an individual scientist may spend a lifetime working in a particular narrow field and never come into contact with the wider context of his or her subject. Indeed, some scientists believe that this is inevitable. For as knowledge accumulates, knowing everything in depth and detail becomes impossible, so that researchers must apparently be content to work in increasingly narrow areas.

Nevertheless, it is still commonly thought to be possible to find examples in which specialization does not lead to fragmentation but rather to an actual extension of the overall context. In biology at the start of this century, for example, most researchers had little to do with the emerging ideas in physics. Experts may have had some superficial knowledge of the new advances in atomic physics and quantum theory but they had little reason to connect it to their daily research. However, a few decades later interest in DNA brought into biology a whole series of new experimental techniques first developed in physics. Today the methods of experimental physics and the quantum theory form an essential part of what has become known as molecular biology. The context of molecular biology has therefore transcended the boundaries of a number of sciences. However, as a new area of study, molecular biology has itself become fragmented and separated from other fields of biology. Today a molecular biologist probably has little in common with workers in whole animal biology, for example. Hence, even when significant cross connections are made between areas of specialization, the end result may in fact be an even more subtle and far-reaching form of rigid specialization.

As was suggested earlier, however, there appears to be no intrinsic reason why the movement between specialization and generalization, analysis and synthesis should of itself necessarily lead to fragmentation. Moreover, it is clear that individual scientists themselves would hardly make a conscious decision to carry out their research in a fragmentary way. How then has the present fragmentation of science come about? Clearly it must involve some particularly subtle factors that have by now become built into the very way that science is carried out. Our proposal is that fragmentation does not so much arise from some defect in the scientific approach. Rather it has its origins in the general ways in which human beings perceive and act, not only as individuals but, more importantly, on an organized social level. As an example (which will be explored in more detail in the next chapter), fragmentation arises in scientific communication and this becomes embedded in the very way the languages of science are used. And since the causes of such fragmentation are in general mainly subliminal, they are extremely difficult to detect and to correct.

A more general subliminal cause of fragmentation in science involves what might be called "the tacit infrastructure of scientific ideas." Some of our most valuable skills exist in the form of such a tacit infrastructure of knowledge. A child, for example, spends long hours with a bicycle before suddenly learning to ride. Yet once this new skill is acquired, it never seems to be forgotten. It takes a subliminal and mainly unconscious form, since no one actually "thinks" about how to ride a bike. Likewise typing, sailing a yacht, walking, swimming, playing tennis, and for the skilled handyperson, fixing a car, replacing a broken electrical plug, or changing a washer in a faucet all involve this sort of tacit infrastructure of knowledge and skills. Similarly, a scientist possesses a great deal of such knowledge and skills which are at his or her "fingertips." These make day-to-day research possible, allowing concentration on the main point of issue without the constant need to think about the details of what is being done. Most scientists, for example, carry out their research by using experimental techniques or applying established theories that were first picked up in graduate school. In this way a physicist may spend a decade investigating, for example, the internal structure of metals without ever needing to question this tacit knowledge in any basic way.

But science, like everything else, is in a constant process of evolution and change. In this process, the developments that are made in one area may sometimes have serious consequences for the foundations of theories and concepts in other areas. In this way, the overall context of science is constantly undergoing changes which, at times, are both deep and subtle.

The result of this complex change is that the underlying tacit infrastructure of concepts and ideas may gradually become inappropriate or even irrelevant. But because scientists are accustomed to using their tacit skills and knowledge in subliminal and unconscious ways, there is a tendency of the mind to hold on to them and to try to go on working in old ways within new contexts. The result is a mixture of confusion and fragmentation.

As an example, consider the development of the theory of relativity. Before Einstein, the Newtonian concepts of absolute space and time had pervaded both the theory and the practice of physics for several centuries. Even a physicist as original as H. Lorentz at the turn of the century continued to use these concepts in an effort to explain the constancy of the velocity of light, irrespective of the speed of the measuring apparatus. Newtonian notions of relative velocity suggested that the measurement of the speed of light should yield an experimental result that depended upon the speed of the observing apparatus relative to the light source. For example, if the apparatus moves rapidly toward the source of light, it would expect to register a higher speed than if it moved away. However, no such effect was observed during very careful measurements. Lorentz, in an effort to retain the Newtonian concepts, proposed an *ether theory*, in which the anomalous results on the measurement of light were explained by actual changes in the measuring apparatus as it moved through the ether.

Lorentz was therefore able to explain the constancy of the velocity of light, independent of the relative speed of the observer, as an artifact produced by the measuring instruments themselves, and there was no need to question the fundamental nature of Newtonian ideas. It took the genius of Einstein to do this. But such was the strength of the tacit infrastructure of basic concepts that it required a long time before scientists could generally appreciate the meaning of Einstein's proposals. As with Lorentz, the general tendency was to hold on to basic ways of thinking in new contexts that called for fundamental changes. In this way a confusion was introduced into the subliminal infrastructure that becomes extremely difficult to detect.

To be free of this confusion, scientists must be able to perceive the underlying infrastructure of skills, concepts, and ideas in a radically new light. In the first instance, such perception reveals various internal contradictions and other inadequacies, which should in themselves alert scientists to the fact that something is going wrong. An accumulation of internal contradictions and inadequacies should properly lead scientists to question the whole general structure of the theories and presuppositions that underlie a particular field. In some cases, this examination would involve calling into question even the independence of that area of specialization from others.

In many cases, however, this sort of response does not actually take place and scientists attempt to press on by putting "new wine in old bottles." But why should this be? The answer to this question involves a psychological factor, the mind's strong tendency to cling to what it finds familiar and to defend itself against what threatens seriously to disturb its overall balance and equilibrium. Unless the perceived rewards are very great, the mind will not willingly explore its unconscious infrastructure of ideas but will prefer to continue in more familiar ways.

The mind's tendency to hold on to what is familiar is enhanced by the fact that the overall tacit infrastructure is inseparably woven into the whole fabric of science as well as into its institutions, on which depends the professional security of each scientist. As a result, there is always a strong pressure against any individual scientist who threatens to "rock the boat." But of course, this resistance is not confined simply to science but occurs in every walk of life when familiar and comfortable thoughts and feelings are threatened. People will therefore tend not to have the necessary energy and courage to call into question the whole tacit infrastructure of their field. But this becomes increasingly difficult to do as the whole infrastructure ultimately extends, in its implications, into the whole of science and even of society itself.

One particularly significant mechanism which the mind employs to defend itself against the inadequacy of its basic ideas is to deny that it is relevant to explore these ideas. Indeed the whole process generally goes further because it is implicitly denied that anything important is being denied! Scientists, for example, may avoid confronting deeper ideas by assuming that each particular difficulty or contradiction can be dealt with through some suitable modification of a commonly accepted theory. Each problem therefore produces a burst of activity in which the scientist seeks a "new idea." But rather than looking for something truly fundamental, scientists often attempt an addition or modification that will simply meet the current problem without seriously disturbing the underlying infrastructure.

Another way of defending the subliminal structure of ideas is to overemphasize the separation between a particular problem and other areas. In this way the problem can be studied in a limited context and without the need to question related concepts. But this only acts to prevent a clear awareness of the ultimate connections of the problem to its wider context and implications. The result is to produce artificial and excessively sharp divisions between different problems and to obscure their connections to wider fields. As these divisions rigidify with time, they cease to constitute valid breaks or abstractions of distinct fields of study and result in a pervasive form of fragmentation. Further work, guided by this fragmentary infrastructure, will lead to an apparent confirmation of the original assumption that there can be a sharp separation between the fields in question. Different areas of study now appear to exist on their own, as objective and independent of the actions, will, and desire of individual scientists, even though their actions originally brought about this fragmentation in the first place. Fragmentation therefore tends to become self-fulfilling, so it is particularly important to become aware of its dangers before being trapped in its consequences.[1]

The resistance of the mind in going beyond the boundaries of particular divisions of subjects, and more generally, its resistance to change in fundamental notions of all kinds, is

1. A fuller discussion of this point is made in author(s) *Wholeness and the Implicate Order*, Routledge and Kegan Paul, London, 1980, Chapter 1.

particularly dangerous where the idea of fundamental truth is involved. Until well into the nineteenth century, most people were willing to believe that humanity, through its common endeavors, was actually approaching some kind of absolute truth about nature. The idea that science could play a significant role in discovering this truth lay behind, for example, the Catholic Church's original reaction to the teachings of Galileo, for it appeared that scientists felt themselves to be in a position to challenge the authority of the Church as the traditional repository of truth. In the nineteenth century, Darwinian evolution produced yet another revolution that appeared to many to strike against the authority of religion.

When science won its battle with the Church for the freedom to entertain its own hypotheses, it in turn became the principal repository of the idea that particular forms of knowledge could either be absolute truths or at least could approach absolute truths. Such a belief in the ultimate power of scientific knowledge evoked strong feelings of comforting security in many people, almost comparable with the feelings experienced by those who have an absolute faith in the truths of religion. Naturally there was an extreme reluctance to question the very foundations upon which the whole basis of this sort of truth rested.

In retrospect, the idea that science can lead to an absolute truth about nature was not initially implausible. After all, in the seventeenth century Galileo and Newton had revealed an impressive internal structure that applied to the whole universe. To many scientists this must have suggested that they were approaching some aspects of the absolute truth. However, science in its ceaseless change soon led to new developments of this "truth" with Darwinism, Freudian analysis, relativity, and quantum theory. Today this process of change has every prospect of continuing. It therefore raises such questions as: How is it possible to reconcile the hope for an absolute truth from science with such radical changes in its very foundations? What is the relationship between scientific ideas and reality if such fundamental changes in scientific theories are constantly taking place? For the modern mind, this notion of absolute truth has become considerably weakened and scientists have become accustomed, at least tacitly, to accepting the need for unending change in their basic concepts. Nevertheless, at the subliminal level at least, most scientists still seem to cling to the hope that in some way the very activity of science will one day bring them to some notion of absolute truth. This appears to be one of the main reasons why they have such a strong disposition to defend the tacit infrastructure of the whole of science with great energy.

Clearly, the whole problem of ending the mind's defense of its tacitly held ideas and assumptions against evidence of their inadequacy cannot be solved within the present climate of scientific research. For within this context, every step that is taken will, from the very outset, be deeply conditioned by the automatic defense of the whole infrastructure. What is needed is some new overall approach, a creative surge of the kind suggested in the Introduction that goes far beyond the tacit and unconscious ideas that have come to dominate science. Such a novel approach would, however, involve questions about the nature of creativity and what, if anything, will help to foster it.

This inquiry begins, in this chapter, by looking at the ways in which creativity has actually operated to give rise to new ideas in physics. This exploration also reveals some of the factors that impede creativity and begins to suggest how the current activities of science could be changed so as to foster a more creative approach. Later chapters take a more general approach as creativity is explored in relation to the whole question of order. Finally, the last chapter explores the implications for a general creative surge, not only in relation to science but also to society and human life as a whole.

NOVELTY AND CONSERVATION IN SCIENTIFIC THEORIES

The ideas discussed in the previous section have something in common with those of Thomas Kuhn, the historian and philosopher of science, whose *The Structure of Scientific Revolutions* aroused much attention in the 1960s.[2] A closer

2. University of Chicago Press, Chicago, 1962.

examination, however, reveals subtle but significant differences between our ideas and those of Kuhn, particularly in connection with the nature of change and conservation in science. More specifically, we differ from Kuhn especially in the interpretation of the breaks that occur in the development of science and in our suggestions of ways in which these can be overcome.

On the basis of a study of the history of how scientific ideas develop, Thomas Kuhn argued that the prevailing activity of science takes place during long periods of what he called "normal science," periods in which the fundamental concepts are not seriously questioned. This "normal science" then gives way to a "scientific revolution" in which theories and ideas change in radical ways as whole new systems of concepts and approaches are created. Kuhn calls these overall systems of concepts and approaches *paradigms*. Paradigms include not only systems of theories, principles, and doctrines, but also what we have called the "tacit infrastructure of ideas" which are transmitted to later generations of scientists in a kind of apprenticeship.

Kuhn argued that, following a scientific revolution, the new paradigm is "incommensurable" with what came before. This clearly suggests a break or fragmentation within the evolution of science. Kuhn's term "incommensurable" is not too clear. It seems to imply that a new paradigm does not have enough to do with the old to permit even a common measure. Incommensurability, in this sense, is quite different from notions such as contradiction or incompatibility, which imply some common infrastructure with opposition lying only in certain points so that a measure of divergence or lack of agreement can be made. Kuhn's term, however, implies that no such measure is possible. Perhaps it would have been better to say that two paradigms are mutually irrelevant. In this sense, those who understand one paradigm might, by a special effort, understand the other. But they would feel that this has little to do with what they regard as the basic framework in which truth is to be grasped. They would have little motive for paying serious attention to the alternative paradigm. Clearly, if Kuhn is correct, the new paradigm therefore leads to a very basic and serious form of fragmentation.

It is therefore important to question Kuhn's whole analysis of this incommensurability and ask if such a fundamental dislocation of ideas must always accompany a scientific revolution. It is also necessary to go into Kuhn's notion of "normal science." In fact, Kuhn did not imply that this was a norm or ideal to which science should approach following a revolution. Rather he argued that this was the traditional way in which scientists have worked in the past. In this book it will be argued that science need not, in fact, work in this way and that Kuhn's "normal science" has no more force than that of custom and habit. A closer analysis will show that during periods of "normal science," quite significant changes, in fact, take place and that true creativity cannot be bound or limited to periods of revolution alone.

But to return to this notion of a scientific revolution. The discussions of this chapter will show that the whole issue is far more subtle than that of opposing two incommensurable paradigms. Indeed there is a potential for a continuously creative approach in science so that any abrupt discontinuity of ideas is not inevitable.

As a preliminary example, let us look at one of the major scientific revolutions to take place before this century, a revolution in the understanding of the basic nature of motion. According to Aristotle, all bodies have their natural place in the universe. When a body has discovered its natural place, it will remain at rest unless acted upon by some external force. Aristotle's theory corresponds to what could be called a "common sense" view of nature. No matter how hard a stone is thrown, experience shows that it eventually comes to rest on the ground. Similarly a cart comes to rest when the horse tires. "Common sense" suggests that all things eventually come to rest, and remain that way unless moved by some external force.

Newton's system, which replaced Aristotle's, argued that the natural state is one of motion, in which rest, or zero velocity, happens to be a special case. An object therefore moves in a straight line, or remains at rest, indefinitely unless some force acts on it. Under the action of a force, the motion changes and its rate of change is expressed by Newton's second law of motion. Newton's laws of motion appear, at first sight, to contradict "common sense," for they suggest that if all opposing forces are somehow removed, then the stone and the cart will continue to move in a straight line forever.

Clearly, the gap between Aristotelian and Newtonian concepts is extremely wide. In the Newtonian paradigm, for example, Aristotle's notion of a natural place is completely beside the point, while in Aristotle's system, there is no room to consider Newton's idea of natural motion. The two ideas, therefore, appear to be unrelated to the point that one is not even relevant to the other. However, a closer examination of Newton's "scientific revolution" shows that, in some areas, a sudden dislocation of concepts and ideas did not in fact take place. To begin with, part of the old tacit infrastructure of ideas was carried over into Newton's scheme. Furthermore, other significant changes occurred in the longer period of "normal science" which followed. In fact, while a considerable confusion of the two sets of ideas and concepts existed for a long period following Newton, this did allow for the possibility of a dialogue between the two paradigms. In this sense the two paradigms were never absolutely incommensurable and indeed some form of creative dialogue could always have been held between them.

During the Middle Ages, people accepted Aristotle's ideas as part of a tacit infrastructure of their worldview. Gradually, following Newton, another worldview arose in which Aristotle's earlier ideas began to seem strange and beside the point. By the end of the eighteenth century, therefore, Aristotle's notions on motion appeared to be quite incommensurable with Newton's. Yet a closer examination shows that, in the Newtonian revolution, not everything had been swept away. For example, the notion of what constituted a material body, and what was the actual structure of space and time, did not change in any radical way, at least at first. The actual Newtonian revolution could best be described, therefore, as a radical change in certain ideas, set against a general background of concepts that remained unchanged.

Indeed the sense of mutual irrelevance of the two paradigms actually arose later, as the implications of Newton's ideas were unfolded across the rest of physics. For as Newton's ideas spread, they began to transform general concepts about the nature of matter that had not been closely examined in the initial "revolution." For example, Newton's laws of motion apply not only to apples and cannonballs but also to the motion of the moon and the planets. But clearly, if the same laws govern the heavens as they do bodies on earth, then there can be no essential difference between these two forms of matter. In this way, one of Aristotle's basic assumptions about the nature of the universe was denied. A similar change took place in the notion of cause, for the approaches of Galileo, Kepler, and Newton showed that the concepts of formal and final causes, advocated by Aristotle, were no longer needed in physics. Therefore, as the new ideas made their effects felt, they began to transform the whole framework of traditional modes of thought. As the old framework was gradually dropped and new notions of cause, motion, and matter evolved, science underwent a major transformation in the way it looked at the world, a change which was absorbed into the new tacit infrastructure.

Clearly, while major changes took place during the Newtonian revolution, the old and new infrastructure of ideas was not, initially, incommensurable on all points. However, in the period of "normal science" which followed, the older infrastructure was eroded and transformed by the many implications of the new ideas. In this way the implications of a "scientific revolution" can be far-reaching and will continue to manifest themselves during the period of "normal science" that follows.

With Einstein's theory of relativity and the quantum theory, traditional concepts of motion, matter, and causality changed yet again. For example, the Newtonian concept of absolute space and time, a holdover from earlier Aristotelian notions, was finally found to be incompatible with Einstein's relativistic ideas. In addition, the classical concept of a particle's trajectory did not cohere with the notion of a continuous quantum transition.

In this and in many other ways, relativity and quantum theory continued the revolution that was started by Newton, and continued to transform the general tacit infrastructure of ideas that underlie physics. One particularly significant aspect of this change was to place a much greater reliance on mathematics. As was pointed out in the Introduction, the idea that mathematics expresses the essential reality of nature was first put explicitly, in modern times, by scientists, such as Sir James Jeans and Werner Heisenberg, but within a few decades, these ideas were being transmitted almost subliminally. As a result, after passing through graduate school, most physicists have come to regard this attitude toward mathematics as being perfectly natural. However, in earlier generations such views would have been regarded as strange and perhaps even a little crazy—at all events irrelevant to a proper scientific view of reality. So yet again, while the scientific revolutions of this century are generally viewed as arising explicitly in the first three decades with quantum and relativity theories, in fact radical changes continued to take place in the postrevolutionary decades that followed. During this period of so-called "normal science," the whole attitude toward the significance of mathematics began to change. Where it had once been regarded as an important tool for unfolding the implications of ideas, concepts, and models, now it was believed to contain the very essence of the scientific ideas themselves.

CREATIVITY AND METAPHORS

Scientific revolutions, therefore, begin with a radical change, which then unfolds, through a long period of "normal" science, into a whole new infrastructure of ideas and tacit assumptions. Of course, such long-term transformations within the largely unconscious infrastructure of ideas, involves the operation of creativity on a continuous basis. In contrast to the approach of Kuhn, therefore, it is argued here that a certain continuity is always preserved during a scientific revolution and that significant changes to this infrastructure continue to take place during the subsequent periods of "normal" science. However, a clear perception of the actual nature of this change does not always take place at the time itself. Indeed, scientists generally believe that "everything changes" in a revolution, while during the longer period that follows, it is assumed that "everything remains basically the same."

In science, this failure to fully perceive the subtle but essential nature of change becomes a major source of rigidity, which in turn contributes to fragmentation in the way that has been described earlier. Hence, in order to understand the essential nature of change in science, it is necessary to see how new ideas actually arise in spite of this rigidity. In doing this, it will be possible to obtain some insight into the nature of creativity, without which science would tend to be caught indefinitely in the "rigid tracks" that it has made through its own progress in the past.

To begin such an inquiry into creativity, consider the example of Newton's theory of universal gravitation. Newton's revolutionary step went far beyond the mere reordering of existing concepts, for it involved a radically new mental perception of nature. The idea that objects may attract each other did not actually originate with Newton. But his genius lay in realizing the full, explicit implications of what was already known within the scientific community. To understand the significance of Newton's perception, it is necessary to go back to the Middle Ages, when science was strongly based on Aristotle's notion that earthly and heavenly matter are of two basically different natures. A great deal of experimental evidence began to accumulate after the Middle Ages which suggested that there is no fundamental difference between heavenly and earthly matter. But this knowledge tended to be kept in one compartment of scientists' minds, fragmented from another compartment which continued to cling to the notion that heaven and earth are separate. Thus scientists never raised the question as to why the moon does not fall, because it seemed evident that, as a result of its celestial nature, it naturally remains in the sky where it belongs.

It was Newton who first perceived the universal implications of the fall of the apple: As the apple falls toward the earth, so does the moon, and so does everything fall toward everything else. To see the universal nature of gravitational attraction, Newton had to become free of the habitual compartmentalization of earthly and celestial matter, a form of fragmentation that was implicit within the tacit infrastructure of the "normal" science of his day. To break away from the habitual and commonly accepted modes of thought, which had been taken for granted for generations, required intense courage, energy, and passion. Newton had these in abundance, and at the height of his powers, he was always asking fundamental questions. The crucial factor in Newton's vision, and indeed in the creation of all new ideas, is this ability to break out of old patterns of thought. Indeed, once this has been done, new perceptions and novel ideas appear to arise naturally.

It is, of course, difficult for the nonscientist to obtain a direct experience of what it is like to create a new theory or scientific concept. But some insight can be gained by thinking about the way metaphors are used in poetry.[3] Shakespeare's plays, for example, are particularly rich in metaphors:

All the world's a stage
And all the men and women on it merely players.
As You Like It

Life's but a walking shadow, a poor player,
That struts and frets his hour upon the stage.
Macbeth

The world *is* a stage; birth and death *are* entrances and exits; life *is* "a tale told by an idiot, full of sound and fury, signifying nothing."

This characteristic use of the word *is* to connect things that are so very different or even incompatible, at first sight, appears to generate a paradox. To express this more explicitly, let A = "the world" and B = "a stage." The corresponding

3. A discussion of metaphor also appears in John Briggs, "Reflectaphors," in Basil Hiley and F. David Peat, eds., *Quantum Implications: Essays in Honour of David Bohm*, Routledge and Kegan Paul, London, 1987.

metaphor then takes the form $A = B$. But common sense dictates that the world is *not* a stage and therefore $A \neq B$. The metaphor therefore appears to involve a simultaneous equating and negating of two ideas, concepts, or objects.

The first sense of the inner significance of a poetic equating of very different things is a kind of tension or vibration in the mind, a high state of energy in which a creative perception of the meaning of the metaphor takes place nonverbally. In some cases this heightened perception is the whole reason for using the metaphor in the first place. Nevertheless some poets chose to go further and show that the two different things that are formally identified are indeed *similar* in some very significant but highly implicit way. In the case of Jacques' speech from *As You Like It*, the meaning of the metaphor between life and a stage is unfolded by comparing people to actors on a stage, and a person's whole life to a series of scenes in a play. Indeed many of Shakespeare's set speeches and sonnets begin with the heightened perception of a metaphor and then, having charged the listener with a high perceptive energy, proceed to unfold the inner meaning of the metaphor by exploring the subtle similarities and differences between A and B.

This notion of a metaphor can serve to illuminate the nature of scientific creativity by equating, in a metaphoric sense, a scientific discovery with a poetic metaphor. For in perceiving a new idea in science, the mind is involved in a similar form of creative perception as when it engages a poetic metaphor. However, in science it is essential to unfold the meaning of the metaphor in even greater and more "literal" detail, while in poetry the metaphor may remain relatively implicit.

These ideas are best explored through an example, Newton's initial insight into the nature of universal gravitation. This can be expressed in metaphoric form as "The moon is an apple," which is then extended to "The moon is an earth." At first, this use of language gives rise to a state of high creative and perceptive energy, which is not basically different from that arising in a poetic metaphor. At this point, therefore, it is sensed that the moon, an apple, and the earth are similar in a very important way, but as with the poetic metaphor, this is not yet expressed explicitly. However, almost immediately, scientific thought realizes that all these objects are basically similar in the sense that they attract each other and obey the same laws of motion. At this stage, while the insight is more explicit, it is still fairly poetic and qualitative in nature. The next step, however, is to transpose the unfolded metaphor into a mathematical language which renders the similarities and differences more explicit. From there on, calculations are used to compare theory with experiment, and explain in detail why all objects fall and yet some, like the moon, never reach the surface of the earth. It should be emphasized here that mathematics has now taken its proper place in the theorizing, for without it, the comparison with experiment could hardly have been made. Moreover, mathematics makes detailed new predictions possible and even leads to new concepts when these ideas are applied in fresh contexts.

A second example of such metaphoric creation is given by the well-known story of how Archimedes was asked to determine the amount of gold in a crown. The philosopher was well aware that if he knew both the weight of the crown and its total volume, he could then calculate its density and determine if this was indeed equal to that of pure gold. If the crown proved to be too light for its particular volume, then Archimedes could conclude that its gold had been adulterated with some other metal. Weighing the crown posed no problem but how was Archimedes to determine its volume? Greek geometry contained a series of rules for working out the volume of various objects, provided that they were of simple, regular shapes. For example, by measuring the length of its sides and applying such a rule, Archimedes could easily have calculated the volume of a cube. But how was he to proceed with such an irregular object as a crown, something that lay outside the whole system of Greek geometry?

Legend has it that Archimedes was resting in his bath when the solution occurred to him. The philosopher observed that the water level in his bath rose as his body sank, and he suddenly equated this process of displacement with the degree to which his body was immersed and then with the volume of another irregularly shaped object—the crown. A metaphor was therefore established between the irregular shape of the crown, the volume of his own body, and the rising water level in the bath. By immersing the crown in water and observing the rise in water level, its volume could therefore be inferred. Archimedes' perception was, to some extent, a visual one, involving the rising of the bathwater. But the essence of his discovery lay in an internal perception of new ideas within the mind, which showed Archimedes how the volume of any object is equal to the volume of water it displaces. The state of high energy and vibrant tension inherent in this instant of creation is captured in the story that at the moment Archimedes saw the key point, he cried out *"Eureka."* Archimedes' perceptive metaphor was later developed in more detail into a general method for the practical determination of irregular volumes and led to the new concept of specific gravity. Finally, with the creation of Newton's calculus, it became possible to place the notion of the volume of an irregular shape on a firm mathematical footing.

Metaphoric perception is, indeed, fundamental to all science and involves bringing together previously incompatible ideas in radically new ways. In *The Act of Creation*, Arthur Koestler explores a similar notion, which he refers to as *bisociation*.[4] Koestler himself makes use of a metaphor between "the logic of laughter" and the creative act, which he defines as "the perceiving of a situation or idea in two self-consistent but habitually incomparable frames of reference." Clearly this is close to what is being suggested here about the role of metaphor.

It is implicit in what Koestler says that creative insights of this kind are not restricted to science, or for that matter the arts and literature, but may arise in any aspect of everyday life. Here the case of Helen Keller, who was taught by Anne Sullivan, is particularly illuminating. When Sullivan came to teach this child, who had been blind and deaf from an early age and was therefore unable to speak, she realized that she would have to give Helen unrestricted love and total attention.

4. Hutchinson, London, 1964.

However, on first seeing her "pupil," she met a "wild animal," who apparently could not be approached in any way at all. If Sullivan had seen Helen only according to "normal," i.e., habitual, perception, she would have given up immediately. But the teacher worked with the child as best she could, with all the energies at her disposal, remaining sensitively observant, "feeling out" the unknown world of the child, and eventually learning how to communicate with her.

The key step was to teach Helen to form a communicable *concept*. This she could never have learned before, because she had not been able to communicate with other people to any significant extent. Sullivan, therefore, caused Helen, as if in a 'game, to come into contact with water in a wide variety of different forms and contexts, each time scratching the word *water* on the palm of her hand. For a long time, Helen did not grasp what all this was about. But suddenly, she realized that all these different experiences referred to one substance in many aspects, which was symbolized by the word *water* on the palm of her hand. A little reflection shows that this is basically similar to the kind of metaphoric perception that was discussed above. Thus, the different experiences were implied in some sense as being equal, by the common experience of the word *water* being scratched on her hand. The perception of the significance of this initiated a fantastic revolution in the whole of Helen's life. Indeed, the discoveries of Anne Sullivan and Helen Keller together were extraordinarily creative in helping to transform not only Helen's life but the lives of a large number of people in similar situations.

It is worthwhile bringing out in more detail just what was involved in this extraordinary act of creative perception. Up to that moment, Helen Keller had perhaps been able to form concepts of some kind, but she could not symbolize them in a way that was communicable and subject to linguistic organization. The constant scratching of the word *water* on her palm, in connection with the many apparently radically different experiences, was suddenly perceived as meaning that, in some fundamental sense, these experiences were essentially the same. To return, for a moment, to the idea of a metaphor, *A* could represent her experience of water standing still in a pail, while B would represent her experience of water flowing out of a pump. As Helen herself said, she initially saw no relationship between these experiences. At this stage, her perception may be put as $A \neq B$. Yet the same word *water* was scratched on her hand in both cases. This puzzled her very much, for it meant that in some way Anne Sullivan wanted to communicate that an equivalence existed between two very different experiences, in other words, that $A = B$. Eventually, Helen suddenly perceived (of course entirely nonverbally, since she had as yet no linguistic terms to express her perception) that A and B were basically similar, in being different forms of the same substance, which was represented symbolically by the word *water* scratched on her palm. At this point, there must have been in Helen a state of vibrant tension, and indeed of intense creative perceptive energy, which was in essence similar to that arising in a poet who is suddenly aware of a new metaphor. However, in the case of Helen Keller, the metaphor did not stop here, but went on to undergo a further rapid unfoldment and development. Thus, as she herself said later, she suddenly realized that *everything has a name*. This too must have been a nonverbal flash of insight because she did not yet have a name for the concept of a name. This perception very probably had its origin in a yet higher order of metaphor, suggested by the fact that Anne Sullivan had been playing a similar "game" with her for many weeks, in which many different "words" had been scratched on her hand, each associated to a number of different but similar experiences. All these experiences were in this way seen to be fundamentally related, in that they were examples of a single yet broader concept, i.e., that of naming things. To Helen, this was an astonishing discovery, for she had in this way perceived the whole general relationship of symbol to concept, starting with water and going on almost immediately to an indefinite variety of things that could be extended without limit.

From here on, the development was more like that of a scientific metaphor than a poetic metaphor, for there was an immense process of unfoldment of the implications of her perceptions in ever more extended form. Thus, she began immediately to learn all sorts of names and to combine them into sentences. Soon she was capable of discourse, along with the thought that goes with it. This power to communicate also opened up her relationship to society, so that she ceased to be a "wild animal" and became a cultured human being.

However, there is another side to this story. Anne Sullivan also worked from a remarkable creative perception. Ordinarily the whole relation of symbol to concept is taken for granted, because it is part of the total infrastructure, which is hardly conscious. Sullivan, however, realized the key importance of scratching the symbol on the palm of Helen Keller's hand to the recurrent feature, while she exposed the girl to many instances of the experience of water. Some essential quality of "waterness" remained constant in all this range of experiences. This perception was so firm and clear that Sullivan worked a very long time against discouraging results before the moment of breakthrough occurred. And when it eventually came, Sullivan was ready to exploit it to the fullest, and immediately bring about a total transformation in the mind of Helen Keller.

It is clear that creative perception in the form of a metaphor can take place not only in poetry and in science but in much broader areas of life. What is essential here is that the act of creative perception in the form of a metaphor is basically similar in all these fields, in that it involves an extremely perceptive state of intense passion and high energy that dissolves the excessively rigidly held assumptions in the tacit infrastructure of commonly accepted knowledge. The differences are in the modes and degrees of unfoldment from the metaphoric to the literal. The experience of Helen Keller and Anne Sullivan begins to show, moreover, that much more than metaphor may be involved in creativity. In their case it was communication, but as we shall see in the succeeding chapters, creativity goes very far beyond even this, into very deep questions of order, both in nature and society. Ultimately, it involves areas that are too subtle for detailed analysis of the kind that is being given here.

HAMILTON-JACOBI THEORY

Fragmentation in science arises in several ways, in particular through the mind's tendency to hang on to what is comfortable and secure in the subliminal infrastructure of its tacit ideas. In turn, this leads to a false perception of the radical nature of change during scientific revolutions, in which old and new paradigms are viewed as "incommensurable," and to an underestimation of the degree of change that takes place during long periods of "normal" science. It is generally believed, for example, that following a major scientific revolution, scientists must consolidate their findings and unfold the new ideas in ways that are not necessarily as creative as in the revolution itself. However, a closer examination of the history of physics shows that a number of very powerful metaphors were developed during this period of "normal" science. Never pursued with sufficient energy and courage to initiate creative new discoveries, they were stifled by the weight of the whole tacit infrastructure of familiar and comfortable ideas that prevailed at the time.

During the nineteenth century, for example, there arose an alternative way of treating the mechanics of moving bodies, which at the time, received too little attention. Newton had approached motion in terms of the definite paths or trajectories taken by particles. These trajectories were calculated using differential equations that were determined by the initial positions and velocities of the particles, and by external forces that acted at each point on the path to produce changes in the particle's speed or direction. The Hamilton-Jacobi theory, developed in the late 1860s, presented a new way of treating motion as based on *waves* rather than on *particles*. In place of treating the motion of a particle as following a given path that is affected by external forces, the Hamilton-Jacobi theory is based upon a wave description in which all motion is perpendicular to a wave front. A simple image is given by the movement of a cork or a small piece of wood that is carried by waves on a lake. In this way the motion is determined by the waves as a whole, rather than by piecewise local actions of a force at each point in the trajectory of the particle.

Clearly, the Hamilton-Jacobi theory is radically different from Newton's. Yet mathematicians were able to show that the two theories, in fact, generated the same numerical results. It was clear that these alternative theories contained essentially "incommensurable" ideas: that matter is in its essence of a particle nature, or that it is of a wave nature. Scientists, however, tended to concentrate upon the *mathematical* aspects of the new theory and to pay less attention to the curious situation that two apparently unrelated notions of the constitution of matter were able to cover the same range of experience and with the same results. In light of the previous section, this situation suggests the possibility of making a metaphorical leap and saying, "A particle *is* a wave."

Such a metaphor that connects the essence of the two theories would have, in a certain sense, anticipated the modern quantum-mechanical notion of wave-particle duality. That is, that the same entity (e.g., an electron) behaves under one set of circumstances as a wave, and in another set of circumstances as a particle. A more detailed discussion and development of this metaphor could have led in the mid-nineteenth century to the general outlines of the modern quantum theory, almost without any further experimental clues at all.

Indeed, William Hamilton had already unfolded the essential similarity between particle and wave, which is implicit in this metaphor, by considering a *ray* of light. This is basically a form of waves and yet has a trajectory resembling that of a particle. The ray, however, still does not quite give a full account of the motion of the particle. To obtain this, it is necessary to consider a wave packet, which consists of a group of waves each having nearly the same wavelength. These waves will combine together within a small region of space to produce an intense disturbance, while outside this region their intensity is negligible. The wave packet therefore suggests a model of a particle that is based on the wave concept. This packet can be shown to move with what is called the *group velocity* and its motion can be calculated from the Hamilton-Jacobi equation. When this is done, the theory yields both Einstein's relationship, which shows that the energy of a "particle" (e.g., a light quantum) is proportional to its frequency, and de Broglie's relation, that the "particle's" momentum is inversely proportional to its wavelength. Both these equations, which are implicitly contained within the Hamilton-Jacobi theory, are in fact the key relationships in the foundations of quantum theory. Furthermore, by means of a small but natural modification to the Hamilton-Jacobi equation, it is possible to derive Schrodinger's equation, the basic equation of the quantum theory.

Scientists in the nineteenth century could, therefore, have been led to most of the essential features of modern quantum theory, through the exploration of the metaphor that "a particle *is* a wave." All that would have been needed to complete the quantum theory would have been to fix certain numerical constants by means of experiment. (Specifically, the numerical constant that appears in de Broglie's relationship. This is implicitly contained within the Hamilton-Jacobi theory and determines the actual values of a particle's frequency and wavelength once a numerical proportionality constant is known. The value of this constant is calculated from a measurement of Planck's constant.)

The essential point which can be learned from this example is that metaphors can sometimes have an extraordinary power, not only to extend the thought processes of science, but also to penetrate into as yet unknown domains of reality, which are in some sense implicit in the metaphor. While all metaphors may not be as powerful as that between the Newtonian and the Hamilton-Jacobi theories, it does suggest that scientific metaphors which link unrelated but fundamental concepts may be especially significant.

In the case of the Hamilton-Jacobi theory, however, scientists were mesmerized by the whole tacit infrastructure of Newtonian mechanics, which viewed motion always in terms of trajectories and particles. By clinging on to this underlying matrix of ideas, it became impossible to treat the wave theory of matter as a potential clue to new fields of reality. Instead the Newtonian theory, along with all its infrastructure of subliminal ideas about particles and trajectories, was believed to correspond to reality in an essential way and the Hamilton-Jacobi theory was treated as an artifact or simply a mathematical transformation of Newton's equations that could be used to facilitate certain types of calculation. So, to the scientists of the nineteenth century, the Hamilton-Jacobi equations did not have any deeper meaning.

Indeed it was a century before scientists were able to seriously consider this new view of reality—a view which had in fact been implicitly contained in what was already known. This move required them to overcome the common presupposition, taken for granted over centuries, that nature is constituted only of bodies that are essentially similar to those of ordinary experience, but simply smaller. One of the major steps in changing this view was taken by Niels Bohr, who proposed, in the spirit of a metaphor, that electrons had discrete orbits that changed discontinuously. Most physicists, however, took this idea in a literal fashion, and for a time, scientific thinking involved an uneasy marriage of classical and quantum concepts. It was not until the radically new perceptions of de Broglie (1924), Heisenberg (1925), and Schrodinger (1926) that physics was able to produce a reasonably complete and consistent theory of the atom. However, if the wave nature of matter, implicit within the Hamilton-Jacobi theory, had been taken as more than a mere mathematical artifact, progress would have been much more rapid. Indeed, the general attitude of "normal" science among nineteenth-century physicists prevented a truly creative extension of classical mechanics and the anticipations of the basic features of the quantum theory.

The Hamilton-Jacobi theory actually contains a number of additional interesting and fruitful metaphors which could also have been explored in the nineteenth century. For example, the equations themselves can be mathematically transformed in a special way so that the actual order of motion they describe is not affected but the wave forms themselves change. In other words, one wave form can be transformed into another wave form without affecting the physical outcome of the motion. What is particularly significant about these "canonical transformations," as they are called, is that in order to leave the actual motion unaffected, time and space must be treated on an equal footing. In other words, time and space become interchangeable, with a new time variable being defined not only in terms of the old time variable but also in terms of the old space variables. The canonical transformation, which are basic to the Hamilton-Jacobi equations, therefore suggest a metaphor in which time and space are, in a certain way, equated. The Hamilton-Jacobi theory therefore anticipates a key feature of both the special and general theories of relativity—that the laws of physics are unchanged (invariant) to transformations in which time and space are treated on an equal footing.

Yet another metaphor can be extracted from the Hamilton-Jacobi theory by considering that these equations can be derived from what is called a *variation principle*. Newton's approach to motion was based on the idea of a differential equation that describes the motion as a succession of steps in which each stage follows continuously from the preceding one in a fashion that is typical of mechanical systems. The variation principle, by contrast, starts not from a *differential equation* (in which the motion is analyzed into infinitesimal steps) but from an *integral* which depends upon the whole trajectory over a finite period of time. This integral is called a *Lagrangian*, and the variation principle approaches motion as if a particle were attempting to "minimize its Lagrangian." The motion of a particle, in this approach, depends upon a whole period of time, rather than, as suggested by Newton, upon a succession of instantaneous mechanical reactions to an external force. The Hamilton-Jacobi theory is, therefore, determined by something that approaches teleology; it appears as if all motion is governed by the need to attain an "end."

Put in the form of a metaphor: Mechanism *is* teleology (of a particular kind). This reverses, in a creative way, the usual habit of thinking in which what appears to be teleology is understood by saying that it can ultimately be explained by, or reduced to, mechanism (of a particular kind). Indeed, if the full implications of this metaphor are unfolded, they could perhaps lead to fruitful new insights into the age-old question as to whether or not there is a teleology in nature and if so, what it means. In a similar fashion, the metaphor in which time and space are equated through the canonical transformations of the Hamilton-Jacobi theory could perhaps lead to deeper insights into the relationship between relativity and quantum theory—an area which at present contains a number of incommensurable features.

Many other examples of the coexistence of very different and perhaps "incommensurable" concepts can be drawn from the history of science. However, in the usual way of doing science, such ideas are not allowed to exist side by side, for one of them generally acts to overthrow the other, along with the earlier paradigm that it represents. Or alternatively, one idea is reduced in force and co-opted or absorbed into the other as a special limiting case or interesting mathematical artifact, so that its deeper meaning is neglected. This was the case with the Hamilton-Jacobi theory. But the various examples already discussed suggest that the actual relationship between concepts may be far subtler and that fruitful insights can flow from equating, in a metaphorical fashion, what at first sight appears to be "incommensurate."

Indeed, a gradual and continuous change along these lines has been taking place over the last decade in elementary particle physics. The first step in this development was the exploration of a kind of metaphor, in which the well-known electromagnetic theory was equated with the very different and, at first sight, almost incommensurable, weak interaction between elementary particles. Through the unfolding of the meaning of this metaphor, considerable progress was made toward bringing order into the theory of the elementary particles. The idea was then extended to introduce the strong interaction in a similar way, and finally attempts were made to bring yet another "incommensurable" force, gravitation, into the system. The ultimate goal of this approach is a *grand unified theory* that would allow all the different forms of physics to unfold from a "symmetrical" ground in which there is no essential difference between them. Further extensions of this approach are now being explored in which the ground is supposed to include newer kinds of force that have yet to be demonstrated experimentally. In this way it may be possible to bring about a solution of the long-standing problems of divergencies and infinities in quantum field theory that have beset theoreticians since the very inception of field theoretic approaches in quantum physics.

SIMILARITIES AND DIFFERENCES: HEISENBERG'S AND SCHRÖDINGER'S APPROACHES TO THE QUANTUM THEORY

The unfolding of a metaphor that equates different and even "incommensurable" concepts can, therefore, be a very fruitful source of insight. But the procedure is by no means straightforward. The first difficulty that is encountered is not only that dissimilar things must be equated but that this must be done in a way that is sensitive to their basic differences, implications, and extensions. Initially, of course, scientists may fail to recognize the essential similarity between different things, for this requires a creative act of perception. But once this perception has been achieved, science may then fail to see the essential differences that are also inherent within the metaphor. Clearly the problem with thought is that it often fails to be perceptively sensitive to similarities and differences and instead applies mechanical habits of seeing similarities and differences.

In the examples given in this chapter it is clear that, in science, perception of similarities and differences takes place primarily through the mind (e.g., Newton's perception of a certain basic similarity between the apple, the moon, and the earth), and much less through the senses. As science developed, this aspect of perception through the mind grew more and more important. Indeed, very little of what could be called direct sense perception takes place in physics today. At one time scientific instruments, such as the telescope and the microscope, could have been considered as direct extensions of the senses, but today, the connection between experimental apparatus and human experience is becoming increasingly remote. The realm of physics is now that of perception through the mind, and theory dominates over experiment in the development of the scientific perception of nature.

An example of the important role that theory plays in scientific perception, and which also shows how perception can fail to make a proper discrimination, is provided by Heisenberg's and Schrödinger's formulations of quantum theory. Initially these were two separate, and almost incommensurate, formalisms that described identical phenomena and could have formed the basis of a potential metaphor. The Heisenberg theory describes atoms in terms of mathematical objects called matrices. These matrices are arrays of numbers that obey well-defined rules of mathematical operation. In Heisenberg's theory, the numbers in these arrays correspond to various observable quantities of the atomic system. While this theory was remarkably successful in accounting for various experimental results on the spectra of atoms (the patterns of light emitted when atoms are excited), it was not able to give a conceptual picture of the atom together with a satisfactory description of what the theory meant. However, within a matter of months after Heisenberg's discovery, Schrödinger published his "wave function" approach, which not only gave correct numerical results but also generated an intuitive picture of the hydrogen atom, in terms of standing electron waves around a central, attractive core.

At first it appeared that Schrödinger's theory would quickly supersede that of Heisenberg and that the latter's theory had been a stopgap procedure on the way to formulating a more complete description of the atom. However, as it turned out, matters were not so straightforward. To begin with, when Schrödinger's equation was extended from the hydrogen atom, with its single electron, to more complicated atomic systems, it became apparent that the wave function itself was by no means as intuitive as had been first supposed. Instead of being a wave in our three-dimensional space, the function existed in an abstract, multidimensional space, and pictures of electron waves around a nucleus proved to be something of an abstraction. But more important, physicists discovered that, under fairly general conditions, the two approaches, Schrödinger's and Heisenberg's, were *mathematically* equivalent.

This formal mathematical equivalence between the two theories was a powerful step but it soon led physicists to ignore the essential differences between the two formulations and to consider them as nothing but alternative methods for calculating results to particular quantum mechanical problems—Heisenberg's approach proving advantageous in some instances and Schrödinger's in others. However, closer analysis shows that the two theories are not *completely* equivalent—an observation which is generally neglected. It turns out that their mathematical equivalence is true only under certain tacit, rather than explicitly stated, assumptions. In fact, more detailed examination of the two approaches shows that they exhibit certain significant differences. In the Schrödinger approach, for example, the quantum system is described by a *wave function*, which is the solution to Schrödinger's equation. Mathematically this wave function is a continuous function in space; in other words, the continuity of space-time is built into the whole Schrödinger theory. By contrast, Heisenberg's theory is not committed to such continuity, for the physical properties of the quantum system do not come from a continuous wave function but from a matrix of discrete numbers.

If physicists had chosen to treat the Heisenberg and Schrödinger theories in a truly metaphoric fashion, then they may also have been able to unfold the implications of their various similarities and differences. Specifically, it would have been possible to explore the idea of space-time both in the context of discreteness and of continuity. In this way, a theory of space-time may evolve which is nonlocal and noncontinuous at very small distances but, in the limit of larger scales, becomes continuous and local. By taking such a metaphorical relationship seriously, and remaining sensitive to both similarities and differences, it may be possible to gain new insights from these two approaches to the quantum theory.

Of course, until the various implications of these theories are formally unfolded, it is not possible to predict if the final results would be important or not. Indeed some physicists would argue that all this is merely "playing" with the formalisms. But creativity always has associated with it an element of play, which may or may not lead to fresh perceptions.

THOUGHT AS PLAY

If science always insists that a new order must be immediately fruitful, or that it has some new predictive power, then creativity will be blocked. New thoughts generally arise with a play of the mind, and the failure to appreciate this is actually one of the major blocks to creativity. Thought is generally considered to be a sober and weighty business. But here it is being suggested that creative play is an essential element in forming new hypotheses and ideas. Indeed, thought which tries to avoid play is in fact playing false with itself. Play, it appears, is of the very essence of thought.

This notion of the falseness that can creep into the play of thought is shown in the etymology of the words *illusion, delusion,* and *collusion,* all of which have as their Latin root *ludere,* "to play." So illusion implies playing false with perception; delusion, playing false with thought; collusion, playing false together in order to support each other's illusions and delusions. When thought plays false, the thinker may occasionally recognize this fact, and express it in the above words. Unfortunately, however, our English language does not have a word for thought which *plays true*. Perhaps this is a reflection of a work ethic which does not consider the importance of play and suggests that work itself is noble while play is, at best, recreational and, at worst, frivolous and nonserious. However, to observe children at play is to realize the serious intensity of their energy and concentration.

Within the act of creative play, fresh perceptions occur which enable a person to *propose* a new idea that can be put forward for exploration. As the implications of this idea are unfolded, they are *composed* or put together with other familiar ideas. Eventually the person *supposes* that these ideas are correct; in other words, he or she makes an assumption or hypothesis and then acts according to the notion that this is the way that things actually are. The movement from *propose* to *compose* to *suppose* enables everyday actions to be carried out with little or no conscious thought. For example, if you *suppose* that a road is level, then you are *disposed* to walk accordingly. After a number of successful trips, you will be further disposed to take it for granted that the supposition or assumption that the road is level is indeed correct, and you will no longer have to think about this point. However, if some part of the road later turns out to be uneven so that you trip, you will be obliged to change your assumption and, through this, a disposition which is no longer appropriate. Taking certain assumptions for granted may be a useful way of freeing the mind to consider other questions, provided it always remains sensitive to evidence that the assumption may, at times, be wrong.

What happens in this relatively simple case may also occur as the mind operates with the theories of science. If, for example, one set of ideas works for a long time, within a particular context, then scientists are disposed to take them for granted and are able to free their minds to focus on other ideas that may be relevant. But this is appropriate only as long as the mind remains sensitive to the possibility that, in new contexts, evidence may arise that shows that these ideas are wrong or confused. If this happens, scientists have to be ready to drop the ideas in question and to go back to the free play of thought, out of which may emerge new ideas.

The above account shows the appropriate relationship between thought and experience. Within such a relationship, creative new perceptions take place when needed as, for example, with the metaphors discussed in this chapter. Such perceptions emerge through the creative play of the mind. It is the very nature of this play that nothing is taken for granted as being absolutely unalterable, and that its outcome and conclusions cannot be known beforehand. In other words, the creative person does not strictly know what he or she is looking for. The whole activity, therefore, is not regarded as a problem that must be solved but simply as play itself. Within this play it is not taken for granted that new things must always be different or that they can never in any significant way be related to what came before. Indeed, it could be suggested that the more different things are, the greater may be the importance in seeing how they are similar, and likewise, the more similar things are, the greater may be the value in perceiving their difference. Science, according to this argument, is properly a continuous ongoing activity. Through creative play and fresh perception there is a constant movement of similarities and differences, with each new theory differing in some subtle but significant fashion from what came before. To sustain this creative activity of the mind, it is necessary to remain sensitive to the ways in which similarities and differences are developing, and not to oversimplify the situation by ignoring them or minimizing their potential importance.

Unfortunately, however, this process, in which experience and knowledge interweave with creative insight, is not generally carried out in the way described above. Indeed it might therefore be called a kind of ideal that is seldom attained or approached. It is not generally carried out because of the common tendency toward unconscious defense of ideas which are of fundamental significance and which are assumed to be necessary to the mind's habitual state of comfortable equilibrium. As a result, there is instead a strong disposition to *impose* familiar ideas, even when there is evidence that they may be false. This, of course, creates the illusion that no fundamental change is required, when in fact the need for such a change may be crucial. If several people are involved, collusion will follow, as they mutually support one another in their false responses.

This often takes place in subtle ways that are extremely difficult to notice. Thus the cases of creative insight discussed earlier all involved becoming aware of certain assumptions that everyone else had, hitherto, taken for granted. Newton's insight into universal gravitation, for example, involved questioning the fundamental difference between earthly and heavenly matter. Indeed since medieval times evidence had been accumulating which should have suggested that heavenly matter and earthly matter were indeed basically similar. So to go on treating the motion of the moon and planets as if it was of a different order from the motion of apples and cannonballs, was, in fact, a false play of thought within the mind. However, the deception involved was a particularly subtle one and most scientists were not consciously aware of its operation. Indeed another form of false play, which enables people to continue in their habitual patterns of thought, is to assume that only a person of considerable genius is capable of a truly creative act. The cases explored in this chapter, however, suggest that genius in fact involves sufficient energy and passion to question assumptions that have been taken for granted over long periods. Of course geniuses must also have the necessary talent and ability to follow through and unfold the implications of their perceptions and questionings. Most people, however, tacitly suppose that they do not have the necessary passion and courage to act in a truly creative way and are doomed to forever "play false" with the more subtle features of their knowledge. They believe that, not being geniuses, they are *restricted* to the tacit infrastructures of subliminally held ideas. But suppose that this assumption is false, and that everyone is potentially capable of truly creative acts in various fields that accord with his or her particular abilities, skills, and knowledge. Clearly a prerequisite for this creativity is that we must cease to take for granted that we are incapable of creativity.

It should now be clear that the mind's disposition to play false in fragmentation and the blockage of free creative play are intimately related. For example, to cling rigidly to familiar ideas is in essence the same as blocking the mind from engaging in creative free play. In turn, it is this very absence of such creative free play that prevents the mind from having the vibrant tension and passionate energy needed to free it from rigidity in the tacit infrastructure of familiar ideas. Indeed, a mind that is forced to cling to what is familiar and that cannot engage in free play is in fact playing false. It has already been compelled to take for granted that it cannot do otherwise. The question of which comes first, the false play or the blocking of free play, like that of the chicken and the egg, is not relevant. They are just two sides of one and the same process.

Closer consideration suggests that it is of the very nature of thought always to engage in some form of play, whether this is free and creative or not. Indeed, even thought that is excessively rigid, and therefore uncreative, is in fact still playing, for it is pretending that certain things are fixed, which in fact are not. Moreover such rigid thought is also at play when it pretends that no pretense is taking place, and that it is being absolutely "serious" and based only on truth and fact. Hence, at the origin of thought, the activity of play cannot be avoided. The only question is whether this play is to be free or false.

It is being suggested in this book that the basic problems of both science and society originate in a general disposition of the mind to engage in a false kind of play, in order to maintain a habitual sense of comfort and security. But this also implies that these problems, at their root, arise through inadequacies in society's current approach to creativity. The great significance of inquiring into the nature of creativity, and what impedes it, is thus evident.

THE HIGH COST OF PARADIGMS—AN ALTERNATIVE VIEW OF SCIENCE AS FUNDAMENTALLY CREATIVE

In order to pursue this whole issue of creativity, it is necessary to return to the question of the nature of paradigms. Paradigms clearly involve, in a key way, the process of taking ideas and concepts for granted, without realizing that this is in fact going on. Since this process takes place as the mind attempts to defend itself against what it believes to be a severe disturbance, a paradigm tends to interfere with that free play of the mind that is essential for creativity. Instead it encourages the process of playing false, especially in deep and subtle areas.

A paradigm, as Kuhn points out, is not simply a particular scientific theory but a whole way of working, thinking, communicating, and perceiving with the mind. It is based largely on the skills and ideas that are tacitly transmitted during what could be called a scientist's apprenticeship, in graduate school for example. However, since the publication of Kuhn's *The Structure of Scientific Revolutions*, many people have equated a paradigm with a fundamental general theory and a change of paradigm with a consciously produced change in this theory. In this way some people go as far as to *propose* a paradigm change. This, however, totally misses the main force of Kuhn's idea, which is that the tacit infrastructure, mostly unconsciously, pervades the whole work and thought of a community of scientists. In this book the original meaning of a paradigm, as proposed by Kuhn, is therefore used.

Up to now in this chapter, a paradigm has been discussed in a negative sense, but it must also be realized that a paradigm has the power to keep a whole community of scientists working on a more or less common area. In a sense, it could be taken as an unconscious or tacit form of consent. At first sight, the paradigm would be of obvious use to the scientific community. However, it also exacts a price in that the mind is kept within certain fixed channels that deepen with time until an individual scientist is no longer aware of his or her limited position. The end result is that each scientist becomes caught in a process of playing false as he or she attempts to maintain this fixed position in situations that call for fundamental change. However, none of this will be apparent to the scientists who work within the paradigm, for they have a common feeling that, within this framework, everything will eventually be solved.

Nevertheless, as time passes, unsolved problems within a given paradigm tend to accumulate and to lead to ever-increasing confusion and conflict. Eventually some scientists, who are generally spoken of as geniuses, propose fundamentally new ideas and a "scientific revolution" results. In turn, these new ideas eventually form the basis of a new paradigm, and sooner or later, this rigidifies into "normal" science. In this way the cycle of revolution and "normal" science continues indefinitely.

Throughout the few centuries of its existence, science has proceeded in this fashion until today it is taken as perfectly normal for revolution to succeed revolution, interspersed by periods of relative stability. But is this whole strategy for doing science inevitable or even desirable? Clearly it results in a degree of confusion and fragmentation which shows little sign of diminishing. Is it possible for science to move in a new direction in which greater freedom for the play of thought is permitted and in which creativity can operate *at all times,* not just during periods of scientific revolution? But if such free play and fundamental creativity were the rule, this would imply that, at any given moment, there would be a number of alternative points of view and theories available in each particular area of science.

Traditionally scientists have assumed that when several theories appear to account for the same phenomenon, then only one of them can be correct. The others are then discarded or, as with the Hamilton-Jacobi theory, a theory will be placed in a subservient position as being useful only in the context of certain calculations. With the Schrödinger and Heisenberg approaches to quantum theory, however, a mathematical transformation connects the two theories and they are therefore taken as saying the same thing. But in other cases, some judgment has to be made. In this case, aesthetics of the mathematics or the logic of the arguments may be used or "Occam's Razor" may be invoked: the injunction of the fourteenth-century philosopher William of Occam that theories with the smallest number of arbitrary assumptions are to be preferred.

There is no logical reason, however, why, in the unfolding of scientific ideas, several theories may not offer alternative but equally valid and important accounts of a particular aspect of nature. Why must some of these theories be rejected almost as a matter of course? It could be objected that if the number of alternative theories became too large, then the whole scientific enterprise would become excessively diffuse and diluted. It is certainly true that without any established limits, ideas do tend to diverge from each other. However, there is also a natural tendency within scientific thinking for ideas to *converge* as well. Intelligent and creative perception of the different theories may, for example, give rise to new metaphors in which ideas are gathered together and the similarities and differences between them are explored and unfolded.

Clearly this tendency, to convergence within divergence, is very different from the sort of convergence that is brought about through a paradigm, in which arbitrary pressures and boundaries are imposed by the, largely unconscious, consensus of the scientific community. Instead it would be as a result of the intelligent perception of the whole situation that a degree of convergence would occur. If science could be practiced in this way, then a more dynamic approach would be possible in which new ideas constantly appear and are then gathered together in creative ways to form limiting cases of yet more general ideas. Within such a dynamic unity there would be an intense motivation toward limiting divergence while, at the same time, avoiding conformity.

This attitude is, in certain ways, similar to that of the late nineteenth-century philosopher William James who advocated a plurality of approaches that are dynamically related. In place of the monolithic unity of the paradigm, which is able to change only by being cracked and shattered in a revolution, would stand a form of unity in plurality.

This proposal, of a creative plurality in scientific ideas and theories, does, however, raise a significant question: What is the relationship of science to reality? Is this plurality simply a matter of developing a number of different points of view which depend on the requirements of society or the particular preferences of the individual? If this is true, then it would appear that the idea of objectivity within science, as a means of obtaining some relative truth about nature, would no longer be valid.

We suggest that there is indeed a meaning to a reality that lies outside ourselves but that it is necessary that we, too, should be included in an essential way as participators in this reality. Our knowledge of the universe is derived from this act of participation which involves ourselves, our senses, the instruments used in experiments, and the ways we communicate and choose to describe nature. This knowledge is therefore both subjective and objective in nature.

It should be emphasized that this approach to reality is very different from that of the logical positivists, a group of mathematicians, physicists, historians, sociologists, and philosophers who began meeting with the philosopher Moritz Schlick in Vienna during the 1920s. The positivists argued that scientific knowledge is essentially a codification of sense data, and they rejected anything that went beyond the direct deduction from sense data as being useless metaphysics. Positivism of this type has had a considerable influence on the thinking of many scientists. The discussions of this chapter make it very clear, however, that the major part of scientific activity is not at all concerned with direct sensation. Much of what could be called "perception" takes place within the mind, in terms of theories: interaction with the external world is mediated through elaborate instruments that have been constructed on the basis of these theories. Moreover, the very questions that science asks arise not from sense data but out of an already existing body of knowledge. So the subjective element in our knowledge of reality comes about not through the senses but through the whole social and mental way that science is carried out.

The essential activity of science consists of thought, which arises in creative perception and is expressed through play. This gives rise to a process in which thought unfolds into provisional knowledge which then moves outward into action and returns as fresh perception and knowledge. This process leads to a continuous adaptation of knowledge which undergoes constant growth, transformation, and extension. Knowledge is therefore not something rigid and fixed that accumulates indefinitely in a steady way but is a continual process of change. Its growth is closer to that of an organism than a data bank. When serious contradictions in knowledge are encountered, it is necessary to return to creative perception and free play, which act to transform existing knowledge. Knowledge, apart from this cycle of activity, has no meaning.

The fact that this knowledge can bring order to experience and even correctly predict new kinds of experience shows that it must in some way be directly related to a reality beyond this knowledge alone. On the other hand, no form of knowledge can be absolutely fixed and apply indefinitely. This means that any search for such absolute, fixed knowledge is illusory, since all knowledge arises out of the shifting, changing activity of creative perception, free play, unfoldment into action, and its return as experience.

But does this mean that truth is a relative affair that depends only on various accidents? Is it possible for society to construct a world in any way it pleases? The answer is that we are, indeed, free to construct knowledge and the world as we desire. However, the results will not always be appropriate, but in some cases may be confused and even destructive. We cannot impose any worldview we like and hope that it will work. The cycle of perception and action cannot be maintained in a totally arbitrary fashion unless we collude to suppress the things we do not wish to see while, at the same time, trying to maintain, at all costs, the things that we desire most in our image of the world. Clearly the cost of supporting such a false vision of reality must eventually be paid.

As a simple example, think of the worldview held by Europeans living in the Middle Ages. This did not include a particularly strong interest in sanitation; indeed sanitation was not very relevant to their worldview. Nevertheless vast numbers of people were killed by plague, in spite of what society happened to believe about the origin and nature of the disease. People did not notice the connection between their suffering and their view, or lack of it, on sanitation. Indeed they probably took it for granted that there could be no such relationship. However, as soon as the true connection was perceived, it became possible to change things in a positive way so that the new worldview led to revolutionary improvements in the prevention of disease and epidemics. The development of this worldview eventually led to the current notion that all disease is related to external causes, such as bacteria and viruses. Disease in the twentieth century is, therefore, considered in terms of causes and cures, a view which is in accord with the general scientific infrastructure of analysis and fragmentation. Only relatively recently have some doctors begun to question the exclusiveness of this current approach and ask: Why is it that, exposed to similar causes, some people catch a disease and others do not? In this way, new perceptions of the nature of disease and the environment, in terms of lifestyle, stress, diet, and neuroimmunology will begin to make themselves felt and may, someday, transform the current view of how it is that people get sick.

Clearly the well-being of society is intimately connected with the particular worldview it happens to hold. It is not simply a matter of "constructing a reality that gratifies us" but of a whole cycle of thought, action, and experience that leads in the long run to the order or disorder of society. As will be seen in more detail in the following chapters, this cycle tends to be blocked, not only during periods of "normal science," when people are insensitive to subtle but important changes, but also during revolutions, when they overemphasize changes and fail to see continuity. Unless a proper sensitivity and clarity about similarities and differences, change and continuity is maintained, rigidity of thought will set in and lead to confusion and inappropriate action—all signs that thought is caught up in "playing false."

FREE PLAY AND POPPER'S NOTION OF FALSIFIABILITY

The proposal of a "unity within diversity," in which the free play of scientific thought permits a number of different theories to coexist in a creative and dynamic fashion certainly goes against the grain of current ideas on how science should operate. The infrastructure of science has been considerably influenced by the ideas of Sir Karl Popper on how scientific theories should be judged. In *The Logic of Scientific Discovery*, Popper points out that a scientific theory cannot be so much proved as be made credible.[5] Repeated experiments, made on the basis of a theory's predictions, will certainly increase its credibility among the scientific community but they can never *prove* its correctness in any absolute sense. All theories are in some way limited, and while a series of experiments may confirm the theory in some limited domain, they cannot rule out the possibilities of exceptions and novel behavior. The best that science can do, therefore, is to falsify a theory, by establishing some significant point of deviation between experiment and prediction.

Popper's ideas produced a significant change in thinking about science, for they showed how understanding begins by provisionally adopting a particular theory, which is confirmed through observations and later falsified and replaced by some

5. Science Editions, New York, 1961.

newer theory. Clearly, in order to be able to acknowledge contradictions between an acceptable scientific theory and actual experience, the theory must ultimately be falsifiable. That is to say, it must be formulated in such a way that its implications are not subject to too many arbitrary assumptions, so that the theory can always be "saved" by suitable adjustment of these assumptions to fit the facts, no matter what these facts turn out to be.

As Popper's ideas permeated into the infrastructure of science, their emphasis was changed somewhat so that today an excessive emphasis is being placed on falsifiability, in the sense that unless a theory can immediately, or very shortly, be compared with an experiment in which it could be falsified, then that theory is not regarded as properly scientific. Without the possibility of some immediate "crucial experiment," the theory is looked on as being "just metaphysics" and without any particular importance for science. The effect of this climate of opinion is to discourage the mind from free play with ideas.

But a new idea which has broad implications may require a long period of gestation before falsifiable inferences can be drawn from it. For example, the atomic hypothesis, first suggested by Democritus twenty-five hundred years ago, had no definitely falsifiable inferences for at least two thousand years. New theories are like growing plants that need to be nurtured and cultured for a time before they are exposed to the risks of the elements.

Rather than putting an exclusive emphasis on falsifiability, it may be more appropriate to suggest that science consists of a two-way movement of confirmation and falsification. Clearly, it makes little sense to go to all the effort of attempting to falsify a hypothesis when it has never actually been confirmed, or if it has little plausibility.

Fundamental ideas need to be sheltered for a while in a spirit of free creative "play." This should be acknowledged within the scientific community as being a necessary period in which the new idea can be discussed openly and refined. Indeed it will be argued in the next chapter that this very

communication is an essential phase in the creative action of
science. If an individual scientist cannot talk about a new idea
seriously until he or she has proposed a definite experiment
that could falsify it, then science will be caught up in a rather
"workaday" attitude in which free play is discouraged unless it
can rapidly be put to the test.

Once, however, a period of nurturing is allowed for a new
theory so that several theories can exist side by side, then the
whole climate of Popper's argument changes. Theories need no
longer be considered as rivals, and the problem of determining
criteria for choosing between them becomes less urgent. It is
even possible that the same scientist may entertain several
alternatives in the mind at once and engage in a free creative
play to see if they can be related, perhaps through a creative
metaphor.

Moreover, in contrast to Popper, it can be argued that a
good general idea that has been falsified in some experiment
could properly be "saved" by a change in its secondary hypotheses. After all, it would be an arbitrary assumption to propose
that this can *never* happen and that all theories come into the
world perfectly formed. Of course, if a scientist got into the
habit of making such adjustments time after time, then this
would suggest that perhaps he or she had been caught up in
"playing false." When the mind is disturbed at the possibility
of having to drop ideas that are dear to it, then it may well
become trapped in subterfuge. On the other hand, when a
person becomes the victim of such behavior, then no method or
philosophical criterion can prevent this from happening. The
mind that wishes to play false will always be able to find a way
around whatever criteria science may decide upon.

The key point here is not therefore to search for a method
that is somehow supposed to prevent scientists from being
caught up in playing false. Rather it is to face the fact that this
whole problem arises because the mind does not wish to
become unduly disturbed. It cannot, in such circumstances,
act creatively but is impelled to play false in order to defend
the ideas to which it has become so attached. What is needed,
therefore, is to press on with this inquiry into the whole nature
of creativity and what impedes its operation.

SUMMARY AND OUTLOOK

To sum up, the current mode of doing science has evolved in such a way that certain of its features seriously discourage creativity. Among these, one of the most important is the development of paradigms. Clearly it is desirable at all times, and not merely during periods of scientific revolution, that there be the possibility of free play of the mind on fundamental questions so that a properly creative response is possible. Paradigms, especially after they have been established for some time, hold the consensual mind in a "rut" requiring a revolution to escape from. Such excessive rigidity amounts to a kind of unconscious collusion, in which scientists unconsciously "play false together" in order to "defend" the currently accepted bases of scientific research against perceptions of their inadequacy.

In this chapter the main form of creativity considered was that of the metaphor. What is essential to this form is that in equating two very different kinds of things, the mind enters a very perceptive state of great energy and passion, in which some of the excessively rigid aspects of the tacit infrastructure are bypassed or dissolved. In science, as in many other fields, such a perception of the basic similarity of two very different things must further unfold in detail and lead to a kind of analogy which becomes ever more literal.

Naturally, not every scientific metaphor will be fruitful any more than every attempt at poetic metaphor is worthy of serious attention. Moreover it is clear that only a person who has gone into a field with great interest and diligence and who has the requisite skills and abilities will be capable of creating a useful metaphor. Even with such people this does not happen very frequently.

Given that the focusing of work in any given field, through the action of a paradigm, gives rise to an excessive rigidity of mind, it was suggested that a better approach is to allow for a plurality of basic concepts, with a constant movement that is aimed at establishing unity between them. Free creative play with ideas would aid in this process and could help scientific thinking to move in fresh and original ways. If this were the case, science would no longer become so rigid that a revolution would be required to bring about basic changes. Indeed this whole process would represent a significant move toward liberating the surge of creativity that is needed if science is to help in confronting the deeper problems of humanity. It is therefore proposed here that such an approach would give rise to a generally better way of doing science than is possible with the traditional approach.

In this chapter, it was shown, through the example of the metaphor, that scientific creativity arises primarily in an act of perception through the mind. In further chapters creativity will be explored in a broader context and there will be no need to focus on the idea of a metaphor and related forms. Thus, in the next chapter, the link of creativity to the act of communication will be explored, and in later chapters, this will be extended to new notions of order.

TWO

SCIENCE AS CREATIVE PERCEPTION-COMMUNICATION

*I*s it possible for science to operate in a radically new way, in which fundamentally different ideas are considered together and new perceptions made between them? In the previous chapter it was shown that the essence of creativity lies in the ability to make such fresh perceptions and it was also hinted that communication plays a key role in such perceptions. In the case of Helen Keller, for example, her moment of insight, and the way in which it was unfolded, involved communication in a very important way. In this chapter the whole question of communication is explored in much greater detail and it is suggested that communication is an essential for the creative act as is perception through the mind. Indeed, within this context, perception and communication are inseparably related, so that creation arises as much in the flow of ideas between people as in the understanding of the individual alone.

PERCEPTION THROUGH THE SENSES AND THROUGH THE MIND

Perception through the senses does not depend upon the immediate physiological details of the eyes or ears alone but on a much wider context that involves the whole disposition of the

63

individual. In the case of vision this has been investigated from a number of different perspectives. Scientists have shown that seeing requires the active movement of both the body and the mind. Visual perception is therefore an intentional and not a passive act.

A clear example of how vision always operates within a wide and general context is given by the case of a person who is born blind and, by means of an operation, is suddenly made able to see. In such instances clear vision is not an instantaneous process, for both the patient and the doctor must first become involved in a great deal of hard work before the meaningless jumble of visual impressions can be integrated into true "seeing." This work involves, for example, exploring the effects of movements of the body on the fresh visual experiences, and learning to relate the visual impressions of an object to the tactile sense that had previously been associated with it. In particular, what the patient has learned in other ways will strongly effect what is seen. The overall disposition of the mind to apprehend objects in particular ways plays an important role in helping to select and give form to what is seen.

These conclusions are confirmed when the nervous system is analyzed at the neurobiological level. In order to see anything at all, it is necessary for the eye to engage in rapid movements which help to extract elements of information from the scene. The ways in which these elements are then built into a whole, consciously perceived picture have been shown to depend strongly on a person's general knowledge and assumptions about the nature of reality. Some striking experiments demonstrate that the flow of information from the higher levels of the brain into its picture-building areas actually exceeds the amount of information that is arriving from the eyes. In other words, what we "see" is as much the product of previous knowledge as it is of incoming visual data.

Sense perception is therefore strongly determined by the overall disposition of both the mind and the body. But, in turn, this disposition is related in a significant way to the whole general culture and social structure. In a similar way, perception through the mind is also governed by these wider issues. A group of people walking through the forest, for example, see and respond to their environment in different ways. The lumberjack sees the forest as a source of wood, the artist as something to paint, the hunter as various forms of cover for game, and the hiker as a natural setting to explore. In each case the wood and the individual trees are perceived in very different ways which depend on the background and expectations of the walker. Clearly the manner in which an overall social disposition influences how things are seen has considerable importance for science. For, as was pointed out in the previous chapter, this mental perception is also linked to the creative act. Clearly the context of creativity extends into a much wider, social field.

It is important, at this stage, to be clear about the exact nature of perception in science. In the seventeenth and eighteenth centuries the human senses generally provided the major source of scientific information. However, by the nineteenth century they began to play a relatively secondary role. In their place, scientific instruments began to supply the primary data of science. During the seventeenth century, relatively simple instruments, such as the microscope and the telescope, could still, however, be regarded as extensions of the eye. But today scientific instruments have grown to such complexity that observations are more and more remote from immediate sense perception.

But of even greater significance is the role of theories, which are now science's major link with reality. Theories determine not only the design of scientific instruments but also the kinds of questions that are posed in the experiments themselves. Clearly, modern scientific instruments can no longer be regarded as simple extensions of the senses. Indeed, even the raw data that they yield are generally fed directly into computers in the form of numbers and digitized signals. In perceiving the external world by means of this computer-processed data, the senses play a particularly minor role when compared with that of thought.

Perception in modern science, particularly in physics, takes place essentially through the mind, and it is here that the inward intention and general disposition most strongly affect what is "seen." For example, the simple intention to look, or the decision to use an object in a certain way, now becomes the intention to investigate the consequences of a theory or the disposition to use certain apparatus.

An additional feature of this scientific perception is its essentially social nature. For without a firm intention shared among many scientists, the complex equipment needed to carry out a modern experiment would never be built and used. The very nature of modern science and its theories is that it gives rise to the design of large and expensive pieces of equipment which require the operation of large institutions. In turn, this predisposes scientists to see nature in a particular way, for it feeds back into their theories and hence into the design of new experiments.

For example, a vast investment on an international scale is currently being made in building and operating elementary particle accelerators. But this, almost subliminally, predisposes scientists to develop theories in terms of particles and to design additional experiments that will give answers in terms of particles again. The whole social structure of physics has the effect of confirming the particle hypothesis of matter. As a consequence, other possibilities become more difficult to investigate.

In stressing that perception in modern science occurs essentially through the mind, it must not be forgotten that this was always a vital component in science. The observational data obtained by Archimedes in his bath, for example, had little value in themselves. What was significant was their meaning as perceived through the mind in an act of creative imagination. The major change occurring in modern science, however, is that this mental perception is more pervasive than it was in earlier times and its social nature is far more dominant.

It should now be clear that all forms of perception—both through the senses and through the mind—involve a cyclic form of activity. Incoming information is apprehended by the mind and, in turn, produces an outgoing activity in which further scanning and information gathering take place in order to confirm, explore, and reinforce what has been seen. This new activity gathers additional information, which is again apprehended by the mind, leading to yet more outgoing activity. But this is very similar to what happens in science as well. Knowledge of reality does not therefore lie in the subject, nor in the object, but in the dynamic flow between them. However, since reality itself is inexhaustible and never fully covered by knowledge, it could also be said to lie outside the subject, while at the same time including this overall cyclic activity.

COMMUNICATION IS ESSENTIAL
TO PERCEPTION IN SCIENCE

Science is essentially a public and social activity. Indeed it is difficult to imagine scientific research, in any real sense, that does not involve communication within the whole scientific community. In other words, communication plays an essential role within the very act of scientific perception. Scientists are disposed in their thinking by a general background, or tacit infrastructure, of ideas, concepts, and knowledge. In addition, they constantly engage in a form of internal dialogue with the whole structure of their particular discipline. In this dialogue a scientist raises questions and meets points of view which are attributed to other scientists and to his or her own past work. In addition to the internal dialogue, scientists are actively engaged in their daily work with a social exchange of ideas and opinions through discussions, lectures, conferences, and published papers. Motivations, questions, and attitudes arise out of these dialogues, so that all scientific research, in the end, arises out of the whole subcultural matrix of science.

When insight occurs, it emerges out of this overall structure of communication and must then be unfolded so that it obtains its full meaning within it. As a particular insight unfolds, the scientist discusses the new ideas with colleagues and eventually publishes them. In this way criticisms are met and new suggestions exchanged within the scientific community. This leads to a transformation of the original perception. This process of general discussion is so pervasive today that it becomes difficult to say who was originally responsible for creating a particular new idea. As each scientist attends seminars, writes papers, and holds discussions with colleagues, new perceptions arise uninvited out of the totality of the social and cultural milieu. Indeed it can truly be said that each scientist contributes something of significance to this communal matrix in which every major scientific discovery has its ultimate ground.

In view of this continuing social flow of ideas, how is it possible for fragmentation to arise to the point where communication becomes seriously blocked? In the previous chapter it was shown how a person can become limited by an overall "infrastructure of ideas" which is held to rigidly and almost unconsciously. But now the danger arises that this structure of ideas not only applies at the individual level but is held by the scientific community as a whole, so that it eventually begins to limit creative acts of perception. It is therefore necessary to make a careful examination of the way communication takes place between scientists. This includes not only individual scientists themselves but the institutions in which research is carried out, and the general attitudes that are fostered and encouraged within the scientific community. Indeed this analysis of communication must be ultimately extended to the whole structure of human relationships themselves. For example, fear and mistrust may be engendered by rigid lines of authority, lack of job security, and concern over status and competition. All these factors conspire to starve that sense of mutual confidence, goodwill, and friendship that is so necessary for the free play and open exchange of ideas.

If science is to engage in a creative new surge, then all this must clearly change. Within this book it is suggested that scientists could engage in a kind of free play of thought that is not restricted by unconsciously determined social pressures and the limitations inherent in particular paradigms. Such free play could be extended into the form of an open dialogue and exchange of ideas within the scientific community so that each scientist becomes more able to realize his or her creative potential. When the tacit infrastructure of thought is no longer held rigidly within the community, then it becomes possible to sustain creativity at a high level throughout the whole of science.

The creative potentialities of free communication are not peculiar to science alone. They were, for example, of crucial importance in the education of Helen Keller, and they can be clearly seen in the operation of the visual arts. Consider a painter who is engaged in making a portrait. A particularly naive view of painting would conclude that the artist is attempting to portray the sitter "as he or she actually is." However, a moment's reflection shows that other artists will portray the same subject in totally different ways. So where does this "artistic truth" lie? An equally naive suggestion is that the artist is primarily concerned with the truth of immediate, "naked" visual perceptions. Yet all sensory data are deeply influenced by a person's background and disposition. In the case of the artist, this includes everything that has gone before in the history of art, as well as with the artist's relationship to the subject.

Psychological experiments have established that visual perception is clearly conditioned by the circumstances in which that perception takes place, for example, the "meaning" of the scene and which questions are put to the viewer at the time. Clearly the artist is not immune to this process and the "artistic vision" arises out of an outward communication with a vast matrix of ideas, social predispositions, and so on. In addition, the artist is also very much concerned with "inward perception," a vision through the mind that is not dissimilar to that experienced by the scientist. These inward perceptions are affected by everything that the painter holds important about the history of art. Indeed the final painting must take its place within an artistic matrix that stretches over space and time. Each painting is an aspect of the history of art and acts to transform and complement it. Manet's "Olympia," for example, owes much to Goya's "The Naked Maja," among other paintings, and, in turn, inspired Cézanne to paint "A Modern Olympia." Throughout the history of art the individual artist's engagement with other painters, sculptors, and poets, and indeed with the whole culture, is intimately tied to the perception and execution of a work.

As in art, so in science does creativity flow out of a free and open communication. Indeed it is not possible to consider any fundamental separation between the mind's perceptions and communication; they are an indivisible whole. Although for the purpose of analysis, it is always possible to divide them into separate parts, in actuality they are two aspects of the same process, which could be indicated by the hyphenated term *perception-communication*. Clearly it is inadequate to think of the scientist as related to reality through individual activity alone. His or her social communication extends throughout the whole scientific community and beyond, for technology acts on the whole society and environment, and in turn, society determines the directions of science through its policies and financial support and in countless other ways. The significance of free and open perception-communication in the creative operation of science makes it of key importance to discover how communication can be blocked or broken and how fragmentation of the scientific endeavor results.

Paradigms and Specialization as Sources of Breaks in Communication

A free-flowing communication is essential to the creative operation of science. However, serious breaks in communication have occurred, particularly within this century, which result in the fragmentary state of science. How do these breaks and barriers to communication come about? One obvious source is the rapid and fundamental changes that have taken place in the development of science. In the movement from Aristotle to Newton, and from Newton to Einstein, new sets of ideas and concepts have appeared which seemed to be irrelevant or incommensurable with older ideas. Indeed, some historians of science have argued that these breaks in communication, and therefore in perception, must always occur during a scientific revolution. We suggest, however, that such a breakdown in communication is not, in fact, inevitable.

Barriers to communication occur not only during revolutions but also in the intervening periods of "normal science." Later in this chapter it will be shown how the special uses of language in science, rather than fostering better communication, in fact act to disrupt the free flow of ideas. A further barrier, and source of fragmentation, is the development of specialized fields of research, for these often include the assumption that ideas and concepts in one field are not really relevant in another.

Of course some degree of specialization is both necessary and desirable. In their day-to-day work, the neurobiologist and the theoretical physicist have little to do with each other's activities. It is not surprising that research into elementary particles or the nature of black holes does not draw upon concepts involving nerve synapses and neurotransmitters. It could hardly be called a serious barrier to communication. The danger arises when it is assumed that, at their deepest levels, these subjects have no true relationship to each other and that the world really does consist of separate parts which can be indefinitely studied on their own. This is the very assumption that underlies fragmentation, and it is worth pointing out, yet again, its basic fallacy. All scientific concepts are founded within a background of ideas that extends across the sciences without limit. Long-range connections between the ideas, approaches, and methods of the various specializations exist that are of crucial importance and cannot be dealt with in terms of separate specializations and disjointed branches within a given field. These long-range connections are often most important when they are subtle and subliminal, so that their influence is indirect. Only when scientific communication takes place in the spirit of creative free play can scientists become sensitive to the overall contexts and long-range connections between their disciplines.

A simple example may illustrate this point. Neurobiologists have little to do with the theories of quantum mechanics. However, it has been found that, in certain ways, the nervous system can respond to individual quanta of energy. This opens the possibility that the current reliance of the neurosciences on everyday notions of space, time, and causality may prove to be inadequate, and eventually notions from quantum theory may have to be brought into this field.

It should also be stressed that each discipline provides a context for the others, contributes to the particular ways they use their scientific language from day to day, and disposes them to perceive nature in particular ways. When to all this is added the constant and often subtle ways in which scientific notions change, it is clear that a constant and active communication is called for. Whenever barriers between the disciplines and specializations become fixed and rigid, then communication breaks down, ideas and contexts become inflexible and limited, and creativity suffers. Indeed the more subtle and unconscious the connections between the sciences, the more dangerous the effect of a blockage to their free flow in active communication.

METAPHORS AS WAYS OF HEALING BREAKS IN COMMUNICATION

Failures in communication within and between the sciences have been shown to be far more subtle and complex than may have at first sight been suspected.

In particular, across the historical context of science, a serious gap in communication occurs between ideas and concepts that are considered to be, using Thomas Kuhn's terms, incommensurable. We suggest, however, that none of these breaks is inevitable and, indeed, that they can be bridged through the creative use of a form of metaphorical thinking.

At a simple level, take, for example, the conflict between the ideas of matter as fundamentally discrete or continuous. Arguments along these lines go back to the ancient Greeks, and at first sight, the two points of view appear to be incommensurable. However, on closer investigation it would appear that any theory of the continuous nature of matter can in fact be based upon an opposing theory involving discrete matter that is so fine as to have never manifested its nature up to the present time. Conversely, any theory of the discontinuous structure of matter can be explained as arising through the localization and concentration of a continuous background.

These two approaches have in fact been explored during this century. Light, for example, which had been thought of as having a continuous nature, was shown to consist of discrete quanta whose size was so small as to have hidden their individual nature until relatively recently. Likewise Einstein proposed that the particulate nature of matter may be explicable as concentrations and knots in a fundamental, continuous field.

In this example, therefore, what at first sight appeared to be incommensurable views, with little communication between them, were, on deeper analysis, shown to have a deeper interconnection. In the previous chapter other metaphors were shown to apply between Newtonian and Hamilton-Jacobian theories of mechanics. A further example is provided by electromagnetic phenomena, which at one time were viewed through two quite separate theories, one dealing with magnetic manifestations and the other with electrical. This example also shows how deeply the theoretical framework affects what is perceived in science.

Eighteenth-century physicists, for example, treated the various manifestations of electromagnetism through two different theories: the theory of magnetism and the theory of electricity. Because particular effects were treated by these different approaches, physicists were never in a position to observe various manifestations as aspects of the one underlying phenomenon. Rather they perceived two quite different classes of events, those arising from magnetic forces and those arising from electrical charges and currents, which were thus fragmented from each other. In a sense the unification of these two fragments of the one whole was first made by J. C. Maxwell with his theory of the electromagnetic field, first formulated in the 1860s. However, it remained to Einstein and his special theory of relativity (1905) to show how a total symmetry can be achieved through the metaphor: electricity *is* magnetism and magnetism *is* electricity.

It is interesting to note that Einstein's seminal paper on the special theory of relativity, *On the Electrodynamics of Moving Bodies,* begins with a consideration of two very different explanations for the one phenomenon—the relative motion of a magnet and an electrical conductor. In one case the magnet is considered to move past the conductor, a loop of wire connected to an electrical meter. Through the electrical field associated with the moving magnet, a current is induced in the wire—the net result is a deflection of the meter. In the second explanation, the electrical conductor is moved past the magnet, which is now at rest. No electrical field is produced in this case; rather the magnetic force on the charged particles (electrons) in the wire cause a current to flow and a deflection of the meter. Two quite different and apparently incompatible explanations are therefore produced for one and the same phenomenon: the flow of an electrical current when a magnet and a wire move relative to each other.

Through his perception that *relative* motion was the essential point, Einstein was led to see electrical and magnetic effects not as absolute and independent but rather as relative to the state of motion. In addition, they are dependent on each other, for an electrical field in one frame of motion is a magnetic field in another frame. What is involved is a kind of metaphor in which electricity and magnetism are equated. Einstein's insight widened the possible range of communication within physics so that today electromagnetic phenomena are perceived in a very different light from what had earlier been the case.

Of course Einstein's perception went beyond this particular case, for he was led to postulate that time is not an absolute. To achieve the new unity between electricity and magnetism, Einstein had to suppose that time, measured in the frame that moves relative to the laboratory (say, the magnet), is different from time measured in the stationary laboratory frame (say, the fixed wire). This laid the basis for a conceptual understanding of what is known as the Lorentz transformation, in which space and time are, in a certain sense, interchangeable.

What emerged from this insight was a new metaphor: time is space. Again two apparently incommensurable concepts were discovered to have a deeper unity, and perception-communication was extended in physics. Indeed Einstein's was one of the most revolutionary steps ever taken in the history of science, which profoundly altered both the mode of communication and the mode of perception of physics.

LANGUAGE IN SCIENCE

The question of communication leads, in a natural way, to a discussion of the whole nature of language in science. Language is normally considered to be a means of communication, but closer analysis shows that it can also lead to particularly subtle, yet deeply significant, breaks in the ability to communicate various concepts between individuals.

The world's languages are almost infinitely rich in their abilities to deal with subtle distinctions of sense and meaning, to the extent that many linguists would argue that whatever can be thought or experienced can be expressed in language. However, in their professional lives people also tend to use language in more specialized and restricted ways. Conversations between lawyers, doctors, or physicists abound in technical jargon, particular turns of phrase, and special usages of language. Indeed within each professional group a particular term will be immediately understood along with all its rich allusions.

Within science, for example, there are even specialized usages confined to fields, such as biology, psychology, chemistry, and physics, in which terms that are basic to one discipline will be irrelevant, different, or even unknown to the other. Such specialized usages of language can of course lead to difficulties in communicating between the disciplines. As a result of the rapidity of its changes, the language of science is constantly changing in subtle but often radical ways. Moreover, as will be shown in this chapter, major changes are not only confined to "scientific revolutions" but may accumulate as a result of the gradual changes that take place during periods of "normal science."

Of course, as with all language use, most of the processes described above take place largely unconsciously and unobserved and reflect the overall infrastructure of ideas embraced subliminally by the scientific community. In this way when fundamental changes in concepts, and in the ways that ideas are actually used, occur, language becomes used in quite new ways while everyone continues to believe that "nothing has fundamentally changed." The result is a serious form of fragmentation in which scientists continue to talk together but in ways that are increasingly at cross-purposes or even incoherent.

This becomes a particular problem in the more mathematical sciences, in which a lack of coherence can occur between the mathematical formalisms and the informal way they are discussed on a day-to-day basis. Indeed a radical change in the way language is used in physics came about as a result of the discoveries of quantum theory. Its implications are so significant that they are worth discussing at length in the next section.

HEISENBERG'S MICROSCOPE EXPERIMENT

In physics before the twentieth century, the meaning of an equation or a concept generally stood in a direct and easily comprehended relationship to something that could be observed or measured. For example, the motion of a particle could easily be identified with the mathematical trajectory given by Newton's laws of motion. According to Newton, the path of a particle is defined once the initial values of its position and its momentum (the particle's mass multiplied by its velocity) are given. The measurement of these two values then enables a physicist to predict the path of a particle.

In the case of a cannonball, a rocket, or an apple, everything is fairly straightforward and there is no lack of coherence between a verbal and a mathematical description. Indeed, the initial values of position and momentum can be measured to a very high degree of accuracy, using, for example, radar, without any appreciable effect on the path of the particle. However, in the case of subatomic particles, the actual probe used to measure position and momentum, a photon of light or a beam of electrons, for example, always perturbs the system in a significant way. The act of measurement therefore has an important effect on what is measured and its full implication in fact led to a radically new use of language in science and to a split between the power of the mathematics, and of the informal language of science, to describe reality.

The reader may already know that the simultaneous measurement of a particle's position and momentum always involves an irreducible degree of quantum mechanical uncertainty. To clarify the implications of this point, Heisenberg devised his hypothetical microscope experiment. The details which follow are, to some extent, unavoidably technical, but they illustrate an important point about the quantum mechanical picture of nature which has far-reaching effects outside science.

The subatomic particle A which is to be measured is located within a target. Suppose that its momentum has already been measured, and to simplify the argument, this is zero: the particle is at rest. The second step is to measure its position which, the reader will anticipate, will be found to involve a degree of quantum mechanical uncertainty. This measurement is done with the help of a fine beam of electrons, E, which hit the particle A. As one of the electrons in the beam hits A, it is scattered and brought into focus by the magnetic lens M so that it falls on the photographic plate at Q and moves on, leaving a track T. By making measurements on this track, something about the particle A can be inferred.

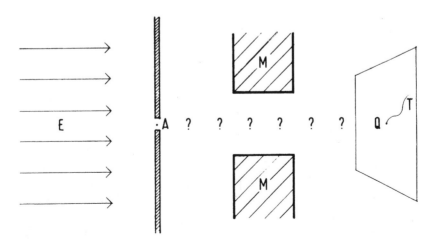

To understand the implications of quantum mechanical uncertainty, it is first necessary to discuss the measurement of momentum as if all the particles obeyed Newton's laws and there were no quantum effects operating. Even in this "classical" case, the colliding electron disturbs the target particle A. Of course by using electrons of very low energy, or by using an electron microscope of very narrow aperture, this disturbance can be reduced as much as desired. But in any case, it is always possible to obtain accurate information about the momentum of A, *even when it is disturbed by the electron beam*. For in the classical case, no fundamental uncertainty is involved. Provided that the structure of the microscope, with its target, magnetic field, and electron beam, is known, then it is always possible to work out the *exact position and momentum* of A by calculating the size of any disturbance made in the measurement.

Hence, although a measurement may involve some finite disturbance, by means of a chain of inferences and using Newton's equations it is always possible to make exact compensations. In this way the notions of trajectory, and of precise values of position and momentum, remain clear, and there is no incompatibility between the mathematical description of a particle's trajectory and the type of informal language used in the above paragraphs.

But to return to Heisenberg's argument, which involves the *quantum mechanical* nature of the link between the particle A and the track T on the photographic plate. In this case the electrons in the beams cannot be considered solely as particles, for they have a wavelike nature as well. The electron can therefore be thought of as a wave which becomes diffracted by A, after which it is focused by the magnetic lens M into a point Q on the photographic emulsion.

In this quantum mechanical case, the appearance of a point Q on the photographic plate can be used only to infer a *probability distribution* of possible points A, from which the electron may have been diffracted. Hence a knowledge of the point Q can be used to calculate the position of the point A to within a certain minimum range of scatter, or uncertainty, ΔX.

A similar argument can be used to infer the momentum of the particle at A. Knowing the direction of the track T gives a probability distribution for the momentum of the particle. Whereas in the "classical" case it was possible to calculate the exact value of any disturbance made during a measurement, in the quantum case this disturbance of momentum can only be known to within a range of scatter, ΔP. Heisenberg was able to show that the total uncertainty involved in this measurement (the uncertainty in position ΔX multiplied by the uncertainty in momentum ΔP) is equal to one of the fundamental constants of nature:

$$\Delta X . \Delta P \simeq h$$

where h is Planck's constant.

Heisenberg was therefore led to conclude that the disturbance made during a quantum mechanical measurement is both unpredictable and uncontrollable, within the limits set by the above uncertainty relation. Indeed this relation is clearly a fundamental principle, inherent in the very nature of reality itself.

Heisenberg's uncertainty relation was of revolutionary signficance in physics. But what was subtler and equally far-reaching was the way in which the informal language of physics (that is, the ordinary descriptive language) had to change. In the discussion above, words, such as *wave, particle, momentum, position, trajectory,* and *uncertainty*, were used, all of which have familiar and well-defined meanings within Newtonian physics. Indeed there is no break or inconsistency between the informal words *trajectory* or *path* of a particle and the mathematical description given by Newton's equations. However, closer analysis shows that these informal terms no longer cohere, in an unambiguous fashion, with the actual mathematical formalism of the quantum theory.

The use of the word *uncertainty* in Heisenberg's analysis of the microscope experiment implies some definite quality whose actual value is not accurately known. If a person is uncertain of the exact time of day because of a faulty clock, this is an expression of ignorance but it does not imply that time itself is uncertain. In a similar case, the way Heisenberg used the word implied that the particle A actually did *possess* a well-defined position and momentum, and therefore a well-defined trajectory, which could not be exactly known, or knowable, to the experimenter. However, the implications of the wave-particle duality of matter, together with the probabilistic relations of quantum theory, are very different. They clearly suggest that the concepts of position, momentum, and trajectory *no longer have any clear meaning*.

Clearly the informal language used by Heisenberg in his original discussion of his uncertainty relationships and the meaning of the formalism itself are totally inconsistent. In other words, a serious gap exists between the way the mathematical formalism is being used and how it is being interpreted. As indicated earlier in this chapter, such a break in communication can lead only to confusion, fragmentation, and the failure of the mind to perceive clearly the nature of reality. Although a more consistent treatment was suggested by Bohr a few months later, nevertheless Heisenberg's initial arguments have been widely accepted by physicists and have entered into the general way the informal language is used. The result is that, almost subliminally, a serious inconsistency has entered into current discussions of the meaning of the quantum theory and pervades much of modern physics. Niels Bohr was able to make a more consistent discussion of Heisenberg's hypothetical experiment by insisting that the precise path of a quantum particle should not be called "uncertain" but, rather, *ambiguous*. That is, something with no clearly defined significance at all. By analogy, consider what is meant by the term *temperature*. Temperature, as measured by a thermometer suspended in the air, is in fact a measure of the mean energy of the air molecules. It is essentially a statistical concept which has a clear definition when a very large number of molecules are involved. But what is the meaning of the temperature of a single molecule, or for that matter of a single atom? Clearly the concept is by no means *uncertain;* rather it is *inherently ambiguous*.

In giving up the notion of a definite, but unknown, position and momentum in favor of inherent ambiguity, Bohr had made a very significant change in the informal way physicists talk about the world. Clearly this approach was more consistent than Heisenberg's, but unfortunately both forms of informal usage have entered the tacit infrastructure of physics, with the result that the language used to talk about reality has become even more confused.

Bohr's arguments were particularly far-reaching and made a far more radical break with classical notions than did those of Heisenberg. In essence Bohr argued that the *entire phenomenon* in which the measurement (or any other quantum measurement, for that matter) takes place cannot be further analyzed into, for example, the observed particle *A*, the incident electron, the microscope, and the plate at which the spot *Q* appears. Rather the *form* of the experimental conditions and the *content* of the experimental results are a *whole* which is not further analyzable in any way at all. In the case of the microscope, this limit to analysis can be clearly seen, for the *meaning* of the results depends upon the way in which the spot *Q* and the particle *A* are linked together. But according to the laws of quantum theory, this involves a single quantum process which is not only indivisible but also unpredictable and uncontrollable.

Bohr made a very careful presentation of this whole question. However, the extreme subtlety of his arguments makes his work relatively inaccessible. The result has been a further degree of confusion in the way physicists communicate with each other, for while most would claim to subscribe to Bohr's position, they do not always realize its full implications. For example, many physicists, if asked whether the electron exists in some fundamental sense, would answer in the affirmative. However, Bohr himself had emphasized that there is no meaning in talking about the existence of the electron except as an aspect of the unanalyzable pattern of phenomena in which its observation takes place. This state of affairs has led to the observation that physicists come to praise Bohr and decry Einstein (because of the latter's refusal to accept the full implications of this feature of the quantum theory) but that they actually think like Einstein while tacitly ignoring Bohr's teaching.

Clearly this state of affairs had led to considerable confusion in the informal language of physics, which makes the quantum theory hard to teach and to discuss. In addition, this situation has been exacerbated by the many other interpretations that have surfaced in the intervening years. For example, the physicist and mathematician von Neumann developed an approach which strongly emphasized the mathematics, logic, and coherence of the formalism.[1] This treatment, which is perhaps closer to the sympathies of physicists than was Bohr's, begins with a set of axioms from which von Neumann attempted to derive the whole subject systematically. While a careful reading of this work shows that the informal language is in fact being used in a new way to discuss the quantum measurement problem, von Neumann would have argued that he was not proposing any changes at all but had simply expressed the essence of the theory within his axioms.

But in discussing a quantum mechanical experiment, von Neumann proposed a sharp conceptual separation between the quantum object and the observing apparatus (the former was described by quantum mechanical laws but the apparatus was assumed to obey Newtonian laws). For von Neumann, the apparatus and the quantum system belong to different worlds, in dynamical interaction with each other. This type of informal discussion is totally incompatible with that of Bohr, who argued that the experimental situation is an unanalyzable whole. For Bohr it would have no meaning to suggest that quantum laws could apply to one part of a system and Newtonian to another.

Following von Neumann, other physicists have added their interpretations. Wigner suggests that the mind of the human observer plays an essential role in a quantum measurement.[2]

1. J. von Neumann, *Mathematical Foundations of Quantum Mechanics*, Princeton University Press, Princeton, 1955.

2. For example, see E. P. Wigner, *Foundations of Physics*, vol. 1, no. 33 (1970); and E. P. Wigner, *Epistemological Perspective on Quantum Theory*, in C. A. Hooker, ed., *Contemporary Research in the Foundations and Philosophy of Quantum Theory*, Reidel, Dordrecht, Holland, and Boston, 1973.

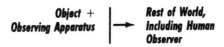

Object + Observing Apparatus	→	Rest of World, Including Human Observer

Von Neumann placed a purely conceptual cut between quantum mechanics and the rest of the world. While the position of the cut was somewhat arbitrary, von Neumann held that quantum theory alone applies to the left of this cut.

Everett argues that the universe, along with its observers, bifurcates each time a measurement is made.[3] The result is a of science in subtly, but significantly, different ways. This has led to a considerable degree of confusion in physics.
proliferation of ideas, each of which uses the informal language

Within this book it is suggested that science will flourish in a more creative way if it allows a diversity of different theories to flourish. When communication between these different points of view is free and open, so that a number of alternatives can be held together at the same time, then it is possible to make new creative perceptions within science. What is proposed is not so much a proliferation of views along with their individual supporters, but rather a *unity in diversity*. This is clearly very different from what has happened in the case of the quantum theory. For many physicists would hold that Bohr, Heisenberg, and von Neumann are all saying the same thing and that there is no essential difference in the content of their approaches and interpretations. However, it should now be clear to the reader that each particular interpretation, in fact, uses the informal language in a radically, but subtly, different way. Yet physicists still tacitly believe that there is no such dislocation in the language.

This example also illustrates the illusion that during periods of "normal science," nothing really changes. For in the decades that followed the revolution of quantum theory, there have been profound changes in the way the theory is to be

3. H. Everett, Jr., *Reviews of Modern Physics*, vol. 29, no. 454 (1957).

understood and interpreted. The resulting confusion, in which language is unconsciously being used in different ways, has given rise to a crisis in communication that makes a clear perception of the quantum theory very difficult.

THE BOHR-EINSTEIN DIALOGUES AND THE BREAKDOWN OF COMMUNICATION IN PHYSICS

The major issue of this chapter is the breakdown in communication within science, especially as it arises in connection with discontinuities between the formal and the informal languages used by scientists. A particularly significant example of this failure to communicate arose between Bohr and Einstein, which, in a symbolic sense, still prevails in physics today.

Bohr first met Einstein during a visit to Berlin in 1920, when the two men discussed the various philosophical issues that lay at the heart of physics. Following Heisenberg's discovery of the uncertainty principle, in 1927, they met at the Fifth Solvay Conference to discuss the meaning of these new discoveries in quantum theory. Throughout the 1930s the two men were involved in a long series of exchanges on the way quantum theory should be interpreted. Bohr, for his part, had introduced new notions into the informal language, so that the meaning of scientific concepts, such as momentum and position, was to be taken as ambiguous. In other words, the meaning of such concepts no longer corresponds in a well-defined way to reality. Einstein, however, believed that fundamental concepts should have, in principle, an unambiguous relationship to reality. This view was in harmony with the essential role that Einstein had assigned to the notion of a *signal* in special relativity. It was a matter of principle that no signal should be transmitted faster than light. But this could not be maintained if the notion of a signal became ambiguous in meaning.

To put this in a different way, both Einstein and Bohr emphasized particular notions of meaning in the informal language of physics. But while, for Bohr, the meaning of fundamental concepts could be ambiguous, in Einstein's view they had to be unambiguous. The two men engaged in a long series of discussions about these issues over the following years. However, in retrospect, it becomes clear that it was never possible to resolve the issues that stood between them because their different uses of informal language implied conflicting notions about the nature of truth and reality and about what is an acceptable type of scientific theory. Bohr began to feel that Einstein had turned in a reactionary way against his own original, revolutionary contributions to relativity and quantum theory. Einstein, for his part, felt that Bohr had become caught in what he called a "tranquillizer philosophy" which avoided fundamental questions. Although the two men had begun as good friends, indeed Einstein said that he had initially felt a real love for Bohr, they eventually drifted apart after many years of fruitless argument and exchanges of challenge and response.

This breakdown between the two men is clearly shown in a story told by Hermann Weyl, who was at the Princeton Institute for Advanced Studies at the same time as Bohr and Einstein. Weyl felt that it was unfortunate that the two men did not get together, so he arranged a party for this purpose. But at the event Bohr and his students congregated at one end of the room and Einstein and his at the other. Clearly the two men had nothing left to say to each other.

Weyl's story shows the enormous power of informal language, which forms a significant part of the tacit infrastructure of science. It also hints at the actual way science is carried out in practice. What is particularly important about this example is the earnest and sustained efforts to maintain contact on what Bohr and Einstein regarded as the key issues in physics. But their differences did not arise within the mathematical formalism itself, for Einstein agreed that the formalism of quantum theory is essentially correct. Yet as a result of the different ways they were using the informal language, the two men became separated by an abyss.

This separation has had particularly serious consequences in the development of relativity and quantum theory, for there is now no common, informal language that covers them both. As a result, although both theories are regarded as fundamental, they exist in an uneasy union with no real way of unifying them. Even within the quantum theory itself there is a serious failure of communication between the various interpretations. Attempts to hold dialogues between these points of view are characterized by the same sort of rigidity, with respect to fundamental assumptions, that was present in the exchanges between Bohr and Einstein. In addition, there is considerable confusion as physicists fail to distinguish between the essential, though extremely subtle, differences between the various approaches.

There is so little awareness of the unbridgeable differences between physicists today because sustained efforts to engage in dialogue have not been made with the kind of persistence shown by Bohr and Einstein. Today the general atmosphere is such that a physicist can do little more than state, and restate, a particular point of view. Various approaches are generally taken to be rivals, with each participant attempting to convince the others of the truth of a particular position, or at least that it deserves serious attention. Yet at the same time, there is a general tendency to regard the whole question of interpretation and the role of informal language as not being particularly important, and instead to focus upon the mathematics about which everyone agrees.

One way of helping to free these serious blocks in communication would be to carry out discussions in a spirit of free dialogue. The nature of such dialogue is discussed in greater detail in Chapter 6, but it seems appropriate to indicate its key features here. What is needed is for each person to be able to hold several points of view, in a sort of active suspension, while treating the ideas of others with something of the care and attention that are given to his or her own. Each participant is not called on to accept or reject particular points of view; rather he or she should attempt to come to an understanding of what they mean. In this way it may be possible to hold a number of different approaches together in the mind with almost equal energy and interest. In this way an internal free dialogue is begun which can lead on to a more open external dialogue. At this stage the mind is able to engage in free play, unimpeded by rigid attachments to particular points of view. It is our suggestion that out of this freely moving dialogue can emerge something that is creatively new, for example, the perception of a new link or metaphor between very different points of view.

It is possible that Bohr and Einstein could have tried to carry out their exchanges in such a spirit. Each could have attempted to discover fundamental points of similarity and difference in what they were saying. Perhaps in this way a creative new metaphor could then have been perceived between their respective points of view. While Bohr and Einstein are now dead, it is still not too late to engage in such a dialogue between the various interpretations of the quantum theory and between quantum and relativity theories. But this will require scientists who are not absolutely committed to particular worldviews. In addition, they would have to give serious attention to the way in which informal language can interfere with the free play of thought that is needed in a creative dialogue between different points of view.

More generally, the opening up of a free and creative communication in all areas of science would constitute a tremendous extension of the scientific approach. Its consequences for humanity would, in the long run, be of incalculable benefit.

THE CAUSAL INTERPRETATION OF THE QUANTUM THEORY

Among the many interpretations of the mathematical formalism of quantum theory is the causal interpretation that was developed by David Bohm over a period of several decades, beginning in the early 1950s. There are several reasons for including a discussion of this theory within this chapter. To begin with, it provides a relatively intelligible and intuitively graspable account of how an actual quantum process may take place. Moreover it does not require a conceptual or formal separation between the quantum system and its surrounding "classical" apparatus. In other words, there is no fundamental "incommensurability" between classical and quantum concepts and, therefore, a greater unity between the formal and informal languages used in its exposition.

In addition, this theory has never before been presented in a nontechnical way and it may be of interest to the reader to learn of a quite novel approach to the quantum theory. In developing the discussion, it is necessary to introduce some new ideas, such as the notion of *active information*, which will become important in explaining the ideas of generative order within Chapter 5. In addition, the causal interpretation has an interesting sociological significance, which arises from the considerable resistance exhibited by physicists to its basic ideas. In the spirit of free dialogue suggested in the previous chapter, the causal interpretation should take its place beside other interpretations in an open exchange of ideas. This may ultimately lead to some creative new perceptions about the nature of physical reality. However, it appears that the largely unconscious commitment to the informal language of the current paradigm of the quantum theory has prevented physicists from responding to this new proposal in any serious way.

THE CAUSAL INTERPRETATION

Although the interpretation is termed *causal*, this should not be taken as implying a form of complete determinism. Indeed it will be shown that this interpretation opens the door for the creative operation of underlying, and yet subtler, levels of reality. The theory begins, in its initial form, by supposing the electron, or any other elementary particle, to be a certain kind of particle which follows a causally determined trajectory. (In the later, second quantized form of the theory, this direct particle picture is abandoned.) Unlike the familiar particles of Newtonian physics, the electron is never separated from a certain quantum field which fundamentally affects it, and exhibits certain novel features. This quantum field satisfies Schrödinger's equation, just as the electromagnetic field satisfies Maxwell's equation. It, too, is therefore causally determined.

Within Newtonian physics, a classical particle moves according to Newton's laws of motion, and the forces that act on the particle are derived from a classical potential V. The basic proposal of the causal interpretation is that, in addition to this classical potential, there also acts a new potential, called the quantum potential Q. Indeed, all the new features of the quantum world are contained within the special features of this quantum potential. The essential difference between classical and quantum behavior, therefore, is the operation of this quantum potential. Indeed, the classical limit of behavior is precisely that for which the effects of Q become negligible.

For the mathematically minded, the quantum potential is given by:

$$Q = \frac{-h^2}{2m} \frac{\nabla^2 |\psi|^2}{|\psi|^2}$$

where Ψ is the quantum field or "wave function" derived from Schrödinger's equation, h is Planck's constant, and m is the mass of the electron or other particle. Clearly the quantum potential is determined by the quantum wave field, or wave function. But what is mathematically significant in the above equation is that this wave function is found in both the numerator and the denominator. The curious effects that spring from this relationship will be pointed out in the following paragraphs.

At first sight, it may appear that to consider the electron as some kind of particle, causally effected by a quantum field, is to return to older, classical ideas which have clearly proved inadequate for understanding the quantum world. However, as the theory develops, this electron turns out not to be a simple, structureless particle but a highly complex entity that is effected by the quantum potential in an extremely subtle way. Indeed the quantum potential is responsible for some novel and highly striking features which imply qualitative new properties of matter that are not contained within the conventional quantum theory.

The fact that Ψ is contained both in the numerator and the denominator for Q means that Q is unchanged when Ψ is multiplied by an arbitrary constant. In other words, the quantum potential Q is independent of the strength, or intensity, of the quantum field but depends only on its *form*. This is a particularly surprising result. In the Newtonian world of pushes and pulls on, for example, a floating object, any effect is always more or less proportional to the strength or size of the wave. But with the quantum potential, the effect is the same for a very large or a very small wave and depends only on its overall shape.

By way of an illustration, think of a ship that sails on automatic pilot, guided by radio waves. The overall effect of the radio waves is independent of their strength and depends only on their form. The essential point is that the ship moves with its own energy but that the *information* within the radio waves is taken up and used to direct the much greater energy of the ship. In the causal interpretation, the electron moves under its own energy, but the information in the *form* of the quantum wave directs the energy of the electron. Clearly the term *causal* is now being used in a very new way from its more familiar sense.

The result is to introduce several new features into the movement of particles. First, it means that a particle that moves in empty space, with no classical forces acting on it whatsoever, still experiences the quantum potential and therefore need not travel uniformly in a straight line. This is a radical departure from Newtonian theory. The quantum potential itself is determined from the quantum wave Ψ, which contains contributions from all other objects in the particle's environment. Since Q does not necessarily fall off with the intensity of the wave, this means that even distant features of the environment can effect the movement in a profound way. As an example, consider the famous double slit experiment. This is generally taken as the key piece of evidence of the wave-particle duality of quantum particles. When electrons are sent through the double slit, they exhibit a wavelike interference pattern on the other side which is quite "incommensurable" with the classical behavior of particles. How does the explanation work in the causal interpretation?

The electron travels toward a screen containing two slits. Clearly it can go through only one slit or the other. But the quantum wave can pass through both. On the outgoing side of

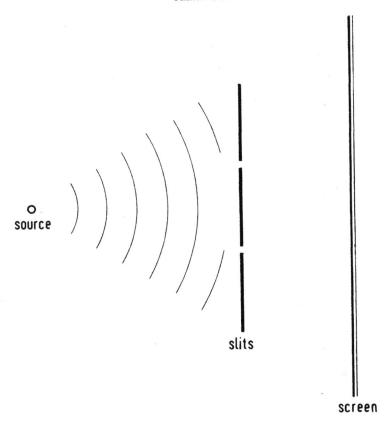

source

slits

screen

The Double Slit Experiment: An electron from the source encounters
the double slits and ends up being registered on the screen behind.
After very many of such individual events a pattern begins to build
up on the screen. The conventional interpretation is that this
interference pattern *is evidence of the wavelike nature of the electron.*
In the causal interpretation, however, the pattern is a direct result
of the complex quantum potential.

the slit system, the quantum waves interfere to produce a
highly complex quantum potential which does not generally fall
off with the distance from the slits. The potential is illustrated
below. Note the deep valleys and broad plateaus. In those
regions where the quantum potential changes rapidly, there is
a strong force on the particle which is deflected, even though
there is no classical force operating. The movement of the

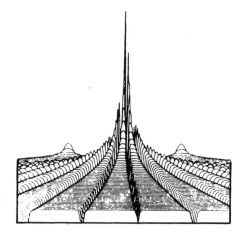

*The quantum potential
for the two slit system*

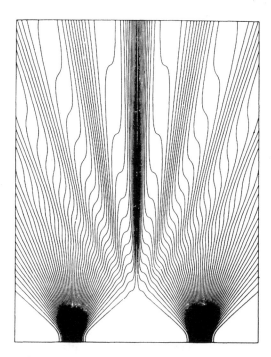

*A collection of trajectories for the electron as it passes through the
two slit system*

electron is therefore modified to produce the scattering pattern shown below. In this case, the wavelike properties do not arise in any essential duality of the quantum particle but from the complex effects of the quantum potential.

The explanation of the quantum properties of the electron given above emphasized how the *form* of the quantum potential can dominate behavior. In other words, information contained within quantum potential will determine the outcome of a quantum process. Indeed it is useful to extend this idea to what could be called *active* information. The basic idea of active information is that a *form*, having very little energy, enters into and directs a much greater energy. This notion of an original energy form acting to "inform," or put form into, a much larger energy has significant applications in many areas beyond quantum theory.

Consider a radio wave, whose form carries a signal—the voice of an announcer, for example. The energy of the sound that is heard from the radio does not in fact come from this wave but from the batteries or power plug. This latter energy is essentially "unformed," but takes up its form from the information within the radio wave. This information is *potentially* active everywhere but only *actually* active when its form enters into the electrical energy of the radio.

The analogy with the causal interpretation is clear. The quantum wave carries "information" and is therefore *potentially* active everywhere, but it is *actually* active only when and where this energy enters into the energy of the particle. But this implies that an electron, or any other elementary particle, has a complex and subtle inner structure that is at least comparable with that of a radio. Clearly this notion goes against the whole tradition of modern physics, which assumes that as matter is analyzed into smaller and smaller parts, its behavior grows more elementary. By contrast, the causal interpretation suggests that nature may be far more subtle and strange than was previously thought.

But this inner complexity of elementary matter is not as implausible as it may appear at first sight. For example, a large crowd of people can be treated by simple statistical laws, whereas individually their behavior is immensely subtler and more complex. Similarly, large masses of matter reduce to simple Newtonian behavior whereas atoms and molecules have a more complex inner structure. And what of the subatomic particles themselves? It is interesting to note that between the shortest distance now measurable in physics (10^{-16} cm) and the shortest distance in which current notions of space-time probably have meaning (10^{-33} cm), there is a vast range of scale in which an immense amount of yet undiscovered structure could be contained. Indeed this range is roughly equal to that which exists between our own size and that of the elementary particles.

A further feature of the causal interpretation is its account of what Bohr called the wholeness of the experimental situation. In, for example, the double slit experiment, each particle responds to information that comes from the entire environment. For while each particle goes through only one of the slits, its motion is fundamentally affected by information coming from both slits. More generally, distant events and structures can strongly affect a particle's trajectory so that any experiment must be considered as a whole. This gives a simple and tangible account of Bohr's wholeness, for since the effects of structures may not fall off with distance, all aspects of the experimental situation must be taken into account.

But if this is the case, then how can the classical world, with its separate and distinct objects, manifest itself? The answer is that in those limits in which the quantum potential can be neglected, its information is no longer active and objects behave as if they were separate and independent. This limit, of negligible quantum potential, is in fact the "classical limit."

In general, this classical limit corresponds to large-scale systems that are not too close to the absolute zero of temperature—in other words, to normal, everyday objects. But there are some phenomena involving low temperatures, such as superconductivity and superfluidity, in which the quantum potential remains appreciable even at the large-scale level. In these cases, quantum effects are significant, as is indeed the case. But the present discussion also opens the possibility of other, large-scale phenomena, as yet undemonstrated, in which quantum effects could manifest themselves.

A significant feature of the causal interpretation is that classical or Newtonian mechanics becomes a special case of quantum mechanics—in which the quantum potential can be neglected. Clearly, the two limits are no longer "incommensurable," but rather one flows naturally out of the other. It shows clearly why Bohr's notions of wholeness must apply in the quantum domain, while in the classical limit the world can generally be analyzed into separate and distinct objects.

In the light of these observations, it would seem reasonable to assume that the causal interpretation should have attracted serious interest and attention among the scientific community. Following critical discussion and, if necessary, modification, the theory would then have assumed its place beside the other interpretations, having its particular adherents and opponents. However, this has not been the case, for the causal interpretation has generally been met with only indifference or hostility. As to why this may be will be explored in the following section.

The theory itself has been worked out in detail and applied to a wide range of examples where it provides a simple and intelligible account of what may be taking place in the quantum domain. Indeed these explanations do not require a break with older, classical concepts, even when quite new concepts are introduced. The result is that the formal and informal languages cohere in a satisfying manner. The formal language, for example, involves equations that deal with particle trajectories and field equations while the informal language speaks of particle paths and fields of information.

A further feature of this interpretation is that it avoids breaks and discontinuities of interpretation. One of the most serious of these, in the conventional approaches, involves what is called the "collapse of the wave function." The behavior of electrons, or other elementary particles, is described conventionally in terms of a wave function (this is formally identical to the quantum wave of the causal interpretation Ψ). According to the Schrödinger equation, this wave function can change only in a smooth and continuous way. However, the results of any quantum mechanical measurement make sense only if it is assumed that the wave function "collapses" in a sudden and discontinuous fashion. Since this collapse is not covered by the Schrödinger equation, and indeed appears to violate it, an additional assumption or some other interpretation is required to explain this "collapse of the wave function." However, in the causal interpretation, the measurement process takes place in a fashion that is entirely natural to the formalism and no additional assumptions are needed.

These considerations are particularly significant when the theory is extended to include cosmology. The origin of the universe in a "big bang" is at one and the same time a cosmological and a quantum mechanical problem. For if energy and matter are created out of a singularity, a vanishingly small region of space-time, then clearly quantum mechanical processes must be involved. But how are these processes and their outcomes to be discussed when the only conceptual framework, in the conventional approach, also involves classical measuring apparatus?

For example, current theories of the origin of the universe rely on what is called "the wave function of the universe." The behavior of this wave function is particularly important at, or near, the moment of origin. But how is this wave function to be properly defined, within the theory, unless classical measuring apparatus is present? Clearly in a cosmological era, in which not even atoms or molecules existed, this apparatus is clearly out of the question. How then is the quantum mechanical origin of the universe to be discussed in any consistent fashion?

Such a problem does not arise in the causal interpretation, for an objective universe can be assumed that is not dependent on measuring apparatus or observers. By contrast, the alternative interpretations either involve a number of extraneous assumptions, or they suppose, along with Bohr, that nothing whatever can be said of the "quantum world." Again, therefore, the informal language coheres with the mathematics and enables an intuitive account to be made of the underlying quantum processes.

A particularly interesting feature of the causal interpretation is that it is not limited to the formal structure of current quantum theory. Its mathematical basis is open to an almost unlimited range of modifications that go far beyond current quantum theory, while still cohering with the informal language of the causal interpretation. All these modifications involve new levels of reality beyond those in which the laws of the current quantum theory are valid. For example, the particle paths described earlier may turn out to be only averages of a more complex set of trajectories (resembling those of Brownian motion) which reflect new levels of reality.[1] These paths (which are discussed in the next chapter) fluctuate chaotically to bring about, in the long run, the same sort of statistical distribution that is predicted in the current quantum mechanics. This shows that the interpretation, while being *causal*, is not strictly *deterministic*. Indeed, in the next chapter it will be shown that the possibility is opened for creativity to operate within a causal framework.

Extensions of the theory involve the introduction of still other types of potentials, in addition to the classical and quantum. There is also the question of an experimental test between the causal and the conventional approaches. A particular extension of the causal interpretation shows that there exists a characteristic "relaxation time," such that if quantum measurements are carried out in short enough time intervals, the statistical results will differ in an important way from those of current quantum theory. It is therefore possible to distinguish experimentally between the predictions of the two theories. However, at present, these relaxation times are too short for existing experimental techniques.

OBJECTIONS TO THE CAUSAL INTERPRETATION

Despite the new features of the theory, and the possibilities for further modifications, the majority of physicists have not considered the causal interpretation to be a serious alternative

1. D. Bohm and J.P. Vigier, *Physical Review*, vol. 96, no. 208 (1954).

to other interpretations of the quantum theory. Why should this be? One of the main reasons is that perhaps it does not cohere with their general way of using the informal language of physics, to which they have become habituated over long periods of application of the usually accepted interpretations. There seems to be little place in their thinking for the causal interpretation, which is therefore ruled out as being irrelevant and not worthy of serious study.

There is also the impression, generally based on a cursory inspection, that the theory is nothing more than a return to older classical concepts which physics has already transcended. Indeed, without a serious study of the approach, physicists will not be properly aware of its genuinely novel conceptual· features and possibilities of as yet unexplored levels of matter.

A particular objection appears to arise from what scientists call the nonlocal nature of the approach. This can be explained in the following way. When several particles are treated in the causal interpretation then, in addition to the conventional classical potential that acts between them, there is a quantum potential which now depends on *all the particles*. Most important this potential does not fall off with the distance between particles, so that even distant particles can be strongly connected. This feature, in which very distant events can have a strong influence, is what is meant by a *nonlocal* interaction and is strongly at variance with the whole spirit of classical mechanics.

There is great reluctance on the part of physicists to consider such nonlocality seriously, even though it does lie at the heart of the formal implications of quantum theory. Because theories in terms of local interactions have been so successful over the past few centuries, the concept is now seen as both necessary and inevitable. But, in fact, there does not ·appear to be any intrinsic reason to rule out nonlocal forces. However, a general attitude has arisen out of the tacit infrastructure of ideas over the last few centuries which prejudices scientists to any serious consideration of nonlocality.

If the objections to nonlocality are based on an almost subliminal prejudice, can they be backed up by physical argument? One suggestion is that nonlocality is inconsistent with the basic tenets of relativity. Nonlocality implies an instantaneous connection between distant events, and this appears to violate the basic principle of relativity that no signal can travel faster than light. However, a more detailed analysis shows that the quantum potential is very "fragile" and unstable to alternations. In other words, if any attempt is made to impose a form on the potential and thereby use it as a signal, this form will become mixed up and lose all order and meaning. The quantum potential cannot therefore be used to carry any signal between distant effects and therefore its instantaneous connection of distant particles does not violate the theory of relativity.

Indeed, there is suggestive evidence that rather than violating physical laws, nonlocality does in fact operate in nature. An experiment initially suggested by Einstein, Rosen, and Podolsky depends upon measuring the nonlocal effects of one distant quantum particle on another. Most recently this experiment has been carried out in Paris by Alan Aspect and interpreted with the aid of a theorem of J. S. Bell. It provides strong evidence for a nonlocal form of interaction. This result follows in a natural way, within the causal interpretation, as a result of the nonlocal quantum potential that directly connects distant particles.

A further objection to the causal interpretation is that it gives only the same predictions as the conventional interpretation. In other words, there is no crucial test between the causal and the usually accepted interpretations. But in fact, the causal interpretation does suggest alternative experimental results, even though they may require refinements in current experimental technology. But more than this, the insistence of a crucial test, or the Popperian criterion of falsifiability, is overrestrictive. A fundamentally new approach of this nature requires a long period of relatively sheltered nurturing before it can properly meet such tests.

Finally there are the objections that certain features of the interpretation are aesthetically unsatisfying. For example, the quantum potential affects the particles but is not affected by them. Within this group, Einstein could perhaps be included and many others like him who feel that the notion of nonlocality is fundamentally unacceptable. On the other hand, if such scientists go as far as to suggest that the very possibility of doing science demands that locality be retained in all its fundamental concepts, then a serious form of demarcation has been set up between science and nonscience.

With the possible exceptions of objections arising from aesthetic judgments, the rest can be summed up by suggesting that the overall informal language of physics, within its present paradigm, is strongly against proposals like that of the causal interpretation. Within this paradigm, or accepted usage of the informal language, must be included Popper's ideas on falsifiability, Heisenberg's absolute requirement that the essence of physics lies in the mathematics, and Einstein's absolute requirement of strict locality. All these factors have, largely unconsciously, come to determine which issues and theories are ruled to be relevant and which irrelevant.

Of particular importance in this subliminal structure is the belief that if several alternative interpretations of the same factual and mathematical content exist, then only one of them can be "right" and the others must be discarded as "wrong." The approach of this book, however, is that science should be carried out in the manner of a creative dialogue in which several points of view can coexist, for a time, with equal intensity. In the case of the Bohr-Einstein exchanges, for example, it was suggested that alternative points of view should have been accommodated without acceptance or rejection. In this way, new, creative insights become possible which constantly emerge out of an open dialogue.

But is such a dialogue possible between the supporters of the currently accepted interpretation and the causal interpretation? As a starting point the causal interpretation could be used whenever a physicist wishes to understand physically what is taking place in a given problem, while one of the other interpretations may be more convenient in carrying out mathematical calculations. Once such a dialogue is established, it becomes possible to go much further and examine the key points of similarity and difference between the different approaches. In this way it may be possible to make creative "metaphors" between time. This, however, requires that the various views be actively held in the mind together, and within the common dialogue of the physics community. For example, the causal interpretation appears similar to Bohr's approach in its emphasis on undivided wholeness, yet different from it, for this wholeness now becomes analyzable in thought. Likewise the causal interpretation is similar to Wigner's in that it gives the "mindlike" quality of active information a primary role, yet different from it, in that it does not imply that the *human* mind can significantly affect the electron in an actual physical measurement.

THE ROLE OF INTERPRETATIONS IN PHYSICS

An interpretation, such as the various interpretations of quantum theory, is in no sense a *deduction* from experimental facts or from the mathematics of a theory. Rather it is a proposal of what the theory might *mean* in a physical and intuitively comprehensive sense. Thus every interpretation brings into the theory something which is not in the observations and equations themselves. This additional material comes from a very broad area which extends beyond what is normally taken to be science and includes philosophy and aesthetic sensibilities.

Bohr's notions of ambiguity and complementarity were, to a large extent, suggested to him by his familiarity with Danish philosophy and the writings of William James with their notion of the "stream of consciousness." Likewise Heisenberg's ideas on the primacy of mathematics did not so much come from his experience in physics but had already been inspired by his familiarity with Plato and, presumably, with the Pythagoreans. Some, however, prefer to take what they believe is a totally pragmatic view and argue that quantum theory is no more than an algorithm for predicting experiments and that to attempt to interpret such an algorithm is a waste of everyone's time. Yet these thinkers, too, have been strongly affected by considerations that lie outside science, such as the opinions of the positivists, operationalists, and empiricist philosophers who were fashionable in the early days of quantum theory.

In essence, *all* the available interpretations of the quantum theory, and indeed of any other physical theory, depend fundamentally on implicit or explicit philosophical assumptions, as well as on assumptions that arise in countless other ways from beyond the field of physics.

Many scientists may find this analysis to be slightly repugnant. Indeed there is a particular ideal today which sees the scientist as "hard-nosed," concerned only with fact and logic and having no time for "soft" content from philosophy or from other "nonscientific" areas. Ernest Rutherford is often held up as the epitome of such a blunt, hardheaded, practical man who had little time for speculation. Such physicists love to quote Rutherford's reply when asked about the new development in quantum theory. "There is only one thing to say about physics: the theorists are on the hind legs and it's up to us to get them down again." But novel hypotheses and daring intuitive leaps into the structure of the nucleus could never have been made by such a cardboard character. The real Rutherford was highly literate, sensitive, and willing to speculate in a bold way as he constructed new experiments and hypotheses.

The image of the "hard-nosed" scientist is yet another example of the subliminal influence that is exerted upon scientists by the tacit infrastructure of ideas of the community at large. Possibly it would be better to regard scientists, in the case of interpretations, as being somewhat like artists who produce quite different paintings of the same sitter. Each theory will be capable of giving a unique insight which is aesthetically satisfying, to a given person, in some ways and not in others. Some interpretations may show creative originality while others may be mediocre. Yet none give the *final* "truth" about the subject.

SUMMARY AND CONCLUSION

The first part of this chapter dealt with the important role played by communication in creative perception, not only in science but also in art. For a variety of reasons, however, this communication tends to break down and so stultifies creativity. An obvious cause of such breaks in communication arises with the development of new paradigms during a scientific revolution. But there is also strong evidence for breaks during periods of "normal" science, which are particularly serious in being largely unrecognized.

The discussion was illustrated with examples of how the formal and the informal aspects of language are currently being used in physics. In particular, the various interpretations of quantum theory and the breakdown in communication between Bohr and Einstein were analyzed.

The second part of this chapter was confined to a detailed discussion of the causal interpretation of the quantum theory, which allows the formal and the informal languages to cohere in a more natural way. This interpretation provides a more intuitive way of talking about events in the microworld. However, partly because this mode of discussion is at variance with that currently adopted in physics, the causal interpretation has been considered irrelevant and largely ignored.

In the next two chapters it will be shown that exploration of the problem of failure in communication and understanding within modern physics requires a thorough analysis of what is meant by order. Indeed it is suggested that the quantum theory demonstrates the need for radically new notions of order, and the confusions and failures associated with theory may be due to an attempt to understand something radically new in terms of an older order—in other words, to "put new wine in old bottles."

THREE

WHAT IS ORDER?

*T*he ultimate failure of Bohr and Einstein to continue their dialogues together symbolizes the degree of fragmentation that exists in physics today. Despite their close friendship and the energy they brought to their encounters, the two men eventually reached the point where they had nothing more to say to each other. In the previous chapter it was suggested that this break in communication was a result of the different and incompatible ways in which the informal language of physics was being used. Each protagonist was using certain terms in particular ways and laying stress on different aspects of the interpretation. A deeper analysis of this whole question shows that what was really at issue was the different notions of order involved. Bohr and Einstein both held to subtly different ideas of what the order of physics, and of nature, should be and this led to an essential break in their dialogue, a break which is reflected in the distance that lies between relativity and the quantum theory even today. In particular, Bohr believed that the order of movement of a particle would admit ambiguity while Einstein felt that such a possibility was too absurd to contemplate. The source of this failure in communication between the two giants of modern physics therefore lay in their incompatible notions of order.

The notion of order extends beyond the confines of a particular theory; permeates the whole infrastructure of concepts, ideas, and values; and enters the very framework in which human thought is understood and action carried out. To understand the full meaning of creativity, and what impedes it, it is necessary to go into the whole nature and significance of order.

The question of order clearly goes beyond the confines of physics, or even science, and into the question of society and human consciousness. Is it possible to inquire into such a vast and general field? Clearly the subject cannot be encompassed in a few pages. The approach that is taken in this chapter, therefore, assumes that the reader has a considerable familiarity with what is meant by order but that this tends to be on an implicit, rather than explicit, level. The subject will therefore be approached in a discursive fashion, as particular aspects, ideas, and intuitions are unfolded. Rather than attempting to make a definition or exhaustive analysis of the nature of order, the intention is to deepen and extend the reader's understanding. This chapter, therefore, focuses on the meaning of order within familiar contexts while the next develops new notions of order that are relevant not only to the ideas of quantum theory but to society, consciousness, and creativity.

New Orders in Society

General notions of order play an incalculably significant role in the totality of human thought and action. When ideas on order change in a really fundamental way, they tend to produce a radical change in the overall order of society. This reaches into every area of life. In fact, the notion of a change in the order of society provides a good starting point for the explorations of this chapter, since it gives some idea of how broad and significant the role of order can be. By examining the order of society it becomes possible to gain a feeling for how subtle and pervasive is the operation of order within the tacit infrastructure of the consciousness of humanity.

The change in the order which underlies society is, in certain ways, not unlike those changes in paradigms that are associated with a scientific revolution. For just as radically new theories are generally taken to be incommensurable with what went before, so new orders of society may arise that are regarded as incompatible with what they replaced. In such cases the whole society is faced with a serious crisis that encompasses everything that was once held dear and is now judged to be irrelevant, improper, or even immoral. In discussing the change of order it is therefore important to ask if all changes in society must necessarily be so destructive and disorienting or if change can happen in more creative ways.

An example of a radical change in the overall order that pervades society can be found in the transition from the Middle Ages to the present day. The medieval worldview is essentially that of a timeless order in which each thing has its proper place, so that even the temporal order of history can be accommodated within the timeless order. This notion can, of course, be traced back to the ancient Greeks, for Aristotle wrote of an eternal order of increasing perfection, going from earthly matter to heavenly matter. An important aspect of this order is that each object has a proper place in the order of things, so that motion of bodies can be understood as a striving to reach this proper place. Within such an order it became natural to view the universe as a single organism.

By the time of the Middle Ages this general scheme had become so comprehensive that it found its eternal basis in the order of religion and philosophy, from which laws, morals, and ethics, which regulated the temporal concerns of society, had their ground. Society as a whole, and each citizen, was regarded as an image of the total, overarching eternal order. Within this framework each individual was able to find a place and a meaning for his or her life. To have a single, encompassing meaning for the universe, society, and the individual was a strong and positive aspect of this eternal order. However, society was not without its share of drawbacks, for the freedoms and rights of individuals were severely restricted and authority was predisposed to become absolute.

After the Middle Ages this order began to give rise to the new secular order in which everything was regarded as being subject to the flux of time. (The "new secular order" is incidentally the motto on the great seal of the United States.) Now nothing had any special space, and motion was reduced to a mechanical process that had no ultimate goal and was therefore going nowhere in particular. The notion of comparing the universe to an organism also gave way to comparing it to a mechanism, and the favorite image of the eighteenth century was that of clockwork.

The secular order was atomistic in nature, and as a result, the individual came to assume a more prominent role in society. This new attitude, of course, helped to bring about an increasing value to human freedom. However, this positive aspect had to be weighed against the negative features. One of the most important of these was that the individual, and indeed the whole society, along with the moral and ethical principles needed for good regulation, no longer had any ultimate meaning. For within the new secular order everything was set in the context of an immense and purposeless mechanical universe, indifferent to human values and to human fate, and in which the earth itself was lost as a mere grain of dust.

All this meant, of course, that a very radical transformation had taken place in the overall order of human life in particular and of society in general. One of the most important aspects of this transformation has been the considerable development of science and technology over the past few centuries. This has helped to dispel the considerable scientific ignorance of the old eternal order and has led to sweeping reforms in medicine and agriculture. However, all these benefits have not been without considerable cost.

Indeed, it is now possible to perceive the dialogue held in the Introduction as concerned with the whole nature of the secular order.

In physics this change of order was especially signaled by Descartes, who introduced the concept of coordinates. These can be thought of as grids by means of which points in space can be located. As the word "coordinate" indicates, they are the basic means by which order is to be described in the new secular and mechanical worldview. Aristotle, for his part, would have understood the formal meaning of Cartesian coordi-

nates, but almost certainly he would have regarded them as
irrelevant to the way he perceived the world. This would have
shown the mutual irrelevance inherent in different notions of
order, which can be thought of as an extension of the mutual
irrelevance of basic ideas in successive paradigms. But the
reader should also keep in mind the possibility of a deeper
continuity between notions of order so that a break in communication between societies does not result.

In Newton's mechanics the order of space and time was
taken as absolute and in this sense, therefore, something
remained from the old Aristotelian order. Within the new
mechanical order was embedded something of the eternal order,
for, according to Newton, space was the "external sensorium of
God" and time flowed equally throughout the universe. However, with Einstein even this remnant of the old order was
called into question. In the theory of relativity, the idea of a time
that flows uniformly across the whole universe was called into
question, for it was shown that the notion of the flow of time
depends on the speed of the observer. No longer could a single
time order span the entire universe; indeed past, present, and
future could not be maintained in the same absolute sense as
for Newton.

With quantum theory, even greater changes in order occurred and the whole basis of the mechanical order, which
formed a significant part of the secular order, was called into
question. It was no longer possible, for example, to define
position and momentum simultaneously, nor could an unambiguous notion of a particle's trajectory be maintained. Indeed by
the third decade of this century the Newtonian order had lost
its clear definition and further new orders were introduced that
depended on the more abstract idea of symmetries, quantum
states, and energy levels.

This whole transformation from the old eternal order has
brought in its wake a movement away from the absolute and
toward the idea that things are inherently relative and dependent on conditions and contexts. But in fact this was the
deeper meaning of giving pride of place to time, rather than
eternity, which originally took place at the end of the Middle

Ages. The essential meaning of time is that everything is mutable and transient. Indeed the Greek god of time, Chronos, swallowed his children. Hence the temporal order is essentially one of change and transience. Admittedly scientists like Newton attempted to formulate universal laws that were assumed to be eternally valid, and therefore were appealing to something that lay beyond time. However, these laws were eventually found to hold only under certain limited conditions and could not be, in this sense, eternal. Even the theory of relativity and the quantum theory, which replaced the Newtonian worldview, are themselves being called into question. The reader will no doubt have heard of "black holes." These are singularities in the fabric of space-time within which all the known laws of physics, including relativity and quantum theory, must break down and basic structures, such as elementary particles, cease to exist. It has even been suggested that the universe itself began in such a "big bang" singularity. Clearly science has reached the point at which everything, in principle, becomes subject to ultimate dissolution within the flux of time. All traces of the eternal order, with its natural cycles and harmonies, have now been swept away.

But these far-reaching changes have not been confined to science alone but have swept into every area of life. In earlier times, for example, people regarded the order of society as eternally determined, perhaps by divine decree. Even though important changes did occur throughout the Middle Ages, for the most part they did not seriously affect those who went through the unchanging and recurring cycles of their lives. However, following the change from eternal to secular order, a series of rapid transformations took place in science and technology, the scale and scope of commerce and industry, the growth of nationalism, and the extension of the general goals of European civilization. For example, the rise of science was followed by a decrease in the importance of religion. Within the scientific order, many religious beliefs appeared to lose their credibility and the progress brought by technology substituted new goals, aims, and values. As the eternal verities and absolute truths became eroded, the older systems of morals and ethics seemed less significant and, in the end, dissolved into a general form of relativism. This change of order even reached into the family as the impersonal ties of commerce, industry, and bureaucracy began to dominate human relationships.

Clearly the change in social order is far-reaching; indeed the social tensions associated with it have not yet been resolved. On the one hand, science and technology have opened up immense new possibilities for a much better life for much of humanity than was ever possible before. On the other, the rise of industry and the growth of technology have given rise to crises in politics, economics, and ecology, and the conflicts between nations have brought us to the brink of nuclear war. Indeed the ever-increasing torrent of change threatens to sweep humanity into a "black hole" singularity. What is inside that singularity is unknown. Will it be increasing misery and ultimate extinction or an unimaginably different and better way of life for all?

In the past, changes in the fundamental order of society have been followed by a period of violence and destruction. This stage of internal conflict and confusion arises when successive notions of order are believed to be incompatible or irrelevant to each other. But is it necessary for a change in order to occur in this way? Is there some intermediate domain in which transition can occur without this associated violence? Is it possible for a range of different notions of order to be held in active suspension within "the mind of society" so that a free dialogue is held between the old and the new orders? In such a case it is possible that an entirely new kind of movement could begin in which the whole society would be in a constant state of creative transformation without disruption.

But up to now, those who have called for major changes in society have given little importance to the question of creativity. Indeed history shows that there has been little conscious realization of what actually takes place during a major change, or where accumulated changes are leading. In general, society changes when a mass of people simply react to particular problems and pressure which have been allowed to accumulate. Even when a few individuals have attempted to confront the issue of change in a creative way they have been hampered by the various issues and problems already brought to light in this book. People, for example, generally tend to be rigidly attached to the tacit infrastructure of their cultural milieu so that they resist all social change in a blind and often destructive way. Others, however, are rigidly attached to the call for revolutionary change and pursue their ends in a similarly blind fashion. Clearly what is called for is a kind of free play within the individual and society so that the mind does not become rigidly committed to a limited set of assumptions, or caught up in confusion and false play. Out of this free play could emerge the true creative potential of a society.

Order and Categories

To understand how a new order can emerge in a creative way, it is necessary to go into the whole notion of order. This will be done by first exploring a relatively detailed idea of order and then generalizing into broader contexts.

Order itself is generally experienced in a number of different situations and contexts. For example, there is the order of number, of points on a line, of space and time, of the movement of a particle through space, and of the functioning of a machine. But order need not be only mechanical or restricted to inanimate systems. There is also the order of growth of an organism, of a language, of thought, of music and art, and of society in general. Indeed it can be truly said that whatever we do presupposes some kind of order. Clearly the subject of order is too broad to be encompassed within an all-inclusive definition. This section, therefore, will begin with the question of how order is thought about, perceived, and brought about in human activity.

Some reflection will show that our first notions of order depend upon our ability to perceive similarities and differences. Indeed there is much evidence which shows that our vision, as well as the other senses, works by selecting similarities and differences. While this can be demonstrated in a number of laboratory experiments and visual illusions, it can be most easily seen through the reader's direct experience. Look around the room for the moment and note how your overall field of vision is particularly sensitive to change and differences of sensation. A sudden small movement is quickly picked up in the corner of the eye. By contrast, the center of the field of vision gives a much finer discrimination of particular forms that are relatively constant. While the background reveals small changes and movements, it is the center of the field which, for example, gives detailed information about a face.

In the Introduction, it was pointed out that damage to the central field still enables meaning to be extracted from the visual field, even if the ability to integrate forms and discriminate fine detail is lost. However, when the background itself is damaged, then information in the central field loses its meaning. This suggests that perception begins through the gathering of differences as the primary data of vision, which are then used to build up similarities. The order of vision proceeds through the perception of differences and the creation of similarities of these differences.

In thought a similar process takes place, beginning first with the formation of categories. This categorizing involves two actions: *selection* and *collection*. According to the common Latin root of these two words, *select* means "to gather apart" and *collect* means "to gather together." Hence categories are formed as certain things are *selected*, through the mental perception of their differences from some general background. To return to vision, an animal may be spotted against the background of the forest or a coin on a patterned carpet may stand out as a result of the glint of its reflection.

The second phase of categorization is that some of the things that have been selected (by virtue of their difference from the background) are collected together by regarding their differences as unimportant while, of course, still regarding their common difference from the background as important. Thus several birds of different size and posture may be abstracted together from the general background of a tree without giving particular attention to the individual differences between them. These birds, however, clearly fall into a different category from any squirrels which are found in the same tree. Categorization therefore involves the combined action of selection and collection.

In the process of observing a flock of birds in the tree the category of birds is formed by putting things together that are simultaneously distinguished from those that do not belong to this category—for example, from squirrels. In this way sets of categories are formed, and these, in turn, influence the ways in which things are selected and collected. Selection and collection therefore become the two, inseparable sides of the one process of categorization.

This determination of similarities and differences can go on indefinitely. For within the similarities of birds will be detected differences between small brownish birds and large black ones. So the category is divided into crows and sparrows, or the new categories of male and female, or perching and flying birds, or birds that sing and birds that are silent are selected. As some differences assume greater importance and others are ignored, as some similarities are singled out and others neglected, the set of categories changes. Indeed the process of categorization is a dynamical activity that is capable of changing in a host of ways as new orders of similarity and difference are selected.

The legends of early humankind, together with contemporary myths from tribes in Africa and North America, suggest that categorization is a primitive but important way of ordering the universe. The gods, for example, are given the task of naming the various animals and plants so as to establish an order in the universe. These legends also indicate that the similarities and differences selected depend upon a context that involves the whole activity and order of the tribe. A people categorize different animals according to their interaction and importance to the life of the tribe. Animals may be selected and identified according to diet, shape and color, habits, or utility. A group of herders in Africa, for example, use a series of words which indicate their sensitivity to variations in the colors of their cattle. In addition, the names of these cattle colors are used to describe other objects. The Inuit (Eskimo) by contrast have quite different priorities for survival and use a number of words to describe the different conditions of ice and snow. Clearly the whole action of categorization is inseparably linked to perception-communication which operates within the overall context of a dynamical social structure.

Most categories are so familiar to us that they are used almost unconsciously. However, from time to time, as the result of some important change in the way we see the world, or as our experience is extended, new categories come into being. Categories are formed which never existed before and new sets of similarities and differences are considered as relevant in entirely new ways. Clearly this implies that perception must be used in a creative way within an ever-changing context.

The creation of new categories relies on a perception that takes place as much in the mind and through the senses. To understand the creative nature of this process, and indeed to develop a theme which will be used throughout this book, the idea of *intelligence* will be introduced. The word *intelligence* is often used in a general and fairly loose way today, but something of its original force can be found in the Latin root *intelligere*, which carries the sense of "to gather in between." It recalls the colloquialism "to read between the lines." In this sense, intelligence is the mind's ability to perceive what lies "in between" and to create new categories. This notion of *intelligence*, which acts as the key creative factor in the formation of new categories, can be contrasted with the *intellect*. The past participle of *intelligere* is in fact *intellect*, which could then be thought of as "what has been gathered." Intellect, therefore, is relatively fixed, for it is based primarily on an already existing scheme of categories. While the intelligence is a dynamic and creative act of perception through the mind, the intellect is something more limited and static. This distinction can be highlighted by suggesting that the IQ test should be more properly said to measure an intellect quotient than an intelligence quotient.

Categories therefore emerge through the free play of the

mind in which new forms are perceived through the creative action of intelligence and are then gradually fixed into systems of categories. But this system of categories always remains fluid and open to further change, provided that the mind itself is open to the creative action of intelligence.

A particularly illuminating example of this whole process is given by the experience of Helen Keller and her teacher Anne Sullivan. When Helen Keller experienced her flash of insight she saw the essential similarity between all the different experiences of water. Anne Sullivan had played a key part in this by helping Helen to select these experiences from the general background and flux of experience, by including them in a kind of game. Helen's moment of insight was the perception of her first category. But this went much further than a simple gathering of basically similar instances, for it had a name that was communicable and which could therefore be used to symbolize the category in thought and elevate it into a concept. But very clearly, Helen's act of perception could not have been based on previous experience, or facts stored in her intellect. It was a pure act of intelligence. Later, however, all this became stored in Helen's memory; it became a part of her tacit infrastructure and a contribution to her intellect.

Categorization can become caught up in exactly those sorts of problems that were discussed in the first two chapters. It is possible for categories to become so fixed a part of the intellect that the mind finally becomes engaged in playing false to support them. Clearly, as contexts change, so do categories. However, when these categories are implicitly embedded in the whole structure of language and society, then they become rigid and persist, in inappropriate ways, within the new context. The result is a form of fragmentation in which significant new connections between categories are ignored, through a false division; and significant differences are ignored within categories, to give a false uniform. Only when the intelligence operates in a free and creative fashion can the mind be free of its attachment to rigid structures of category and is then able to engage in the formation of new orders.

A FORMAL REPRESENTATION OF ORDER

The generation of categories is one aspect of the formation of order but it does not go far enough. While it allows for an infinite variety of sets of categories that depend on general contexts, it is not yet sufficiently developed and self-determined. However, a more definite scheme can be accomplished by applying the notions of similarities and differences to themselves in a series of levels.

It is therefore proposed that a particular general notion of order can be understood in terms of similar differences and different similarities. Consider the example of a line. It can be thought of as characterized, or indeed constructed, out of a series of equal segments in contact: *a*, *b*, *c*, *d*, *e*, *f*, etc. The characteristic of the line is that the difference between *a* and *b* is similar, and indeed equal to, the difference between *b* and *c*, and between *c* and *d*, and so on.

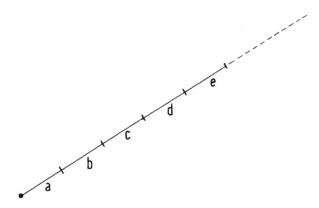

The order of the line is defined by a single, similar difference.

In a similar way, it is possible to analyze a curve, such as a circle, by approximating it to a polygon of many sides. The difference between the segments now includes not only their lengths, which are all equal, but also their angles, which are also equal. Again, the difference between *a* and *b* is similar to the difference between *b* and *c*, *c* and *d*, and so on. The circle is therefore defined by a single, similar difference.

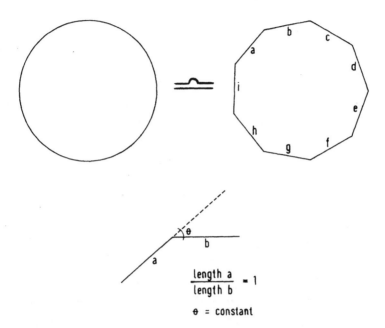

$$\frac{\text{length a}}{\text{length b}} = 1$$

θ = constant

When the lengths of segments progressively diminish in such a way that the difference between successive segments is still similar, then a spiral results.

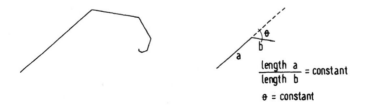

$$\frac{\text{length a}}{\text{length b}} = \text{constant}$$

θ = constant

In three dimensions, the line may turn outside the plane of the spiral and a series of similar differences will give rise to a helix.

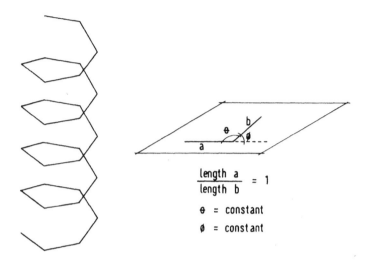

$$\frac{\text{length a}}{\text{length b}} = 1$$

θ = constant

ϕ = constant

In this fashion, a whole class of curves can be constructed in which the differences between neighboring segments are all similar. While the examples actually illustrated so far have been quite simple, by making the constant difference as complex as desired, the geometry of the curves can be enriched.

Even the trajectories of Newtonian mechanics are determined in this way. Newton's first law indicates that the natural motion is one in which all segments are equal to each other— straight line motion or rest. The second law shows that the rate of change of this motion is equal to the external force. In the case of a constant force, such as gravity, this indicates that the differences between successive small segments of the velocity are all similar, and indeed equal.

This concept of order therefore makes it possible to fully appreciate Newton's metaphor about the apple and the moon: the order of similar differences in the motion of the falling apple is similar to the order of similar differences within the orbit of the moon. Newtonian mechanics, operating with constant external forces, is therefore encompassed within the definition of order through similar differences. An extension of the scheme enables more complex cases of motion to be treated as well.

CONSTITUTIVE AND DESCRIPTIVE ORDER

Before going on to discuss more complicated cases of order, such as chance and chaos, it is important to make a distinction between what could be called constitutive order and descriptive order. Consider, for example, the flight of an aircraft in terms of its coordinates on a map. Clearly this involves a descriptive order because the coordinates themselves do not have any material existence of their own with respect to the aircraft. In a similar way, an architect's plans for a house are also a form of descriptive order. However, in discussing the construction of a beehive in terms of individual hexagons, or a wall in terms of bricks, this clearly involves the very constitution of the object in question. Such orders will be called constitutive.

But it is equally true that the discussion of the order of a geometric curve or a trajectory involves both a descriptive *and* a constitutive order, in the sense that the latter order constitutes the very essence of the figure in question. Thus the spiral can be described in terms of a series of segments having a similar difference, but it is equally true that the spiral is actually built out of such segments. Indeed the distinction between descriptive order and constitutive order is never absolute, for every constitutive order has some descriptive significance and every descriptive order has a constitutive basis. For example, repeating hexagons are a convenient way of describing a beehive as well as for constructing one. Likewise the architect's plans have a constitutive basis in paper and ink. Of course this basis has very little relevance to the order of the house. However, once it is converted to marks on lengths of wood, plumb lines, and scaffolding, it begins to lie midway between a constitutive and a descriptive order. An additional example is given by the painter who uses a coordinate grid to enlarge a painting or to transfer a cartoon onto a wall. Using a series of rows and columns of pinpricks, the coordinate system becomes intimately connected to the constitutive order of the final work.

These discussions lead naturally to a particularly important question: Is order simply within the mind? Or does it have an objective reality of its own? In examining the symmetry of a snowflake, starfish, sunflower, and snail's shell, it seems clear that a particular, simple form of order is of the very essence of the object's form. But what of subtler forms of order, such as vortices and emerging structures? (These are described in the next chapter.) What meaning can be ascribed to statements like "the elementary particles are ordered according to an SU(5) symmetry" or "the order of the universe arose through the 'breaking' of a certain symmetry"? What is the underlying meaning of Lévi-Strauss's claim that primitive societies are ordered on the basis of internal structures that are "not without similarity to Boolean algebra," or the current notion of biology that the life of the organism is based on the order of information within its DNA, or psychologist Carl Jung's assertion that the order of the psyche, and indeed that of the universe as a whole, has its ground in certain archetypes? To what extent are these orders and symmetries simply functions of the human mind and to what extent do they have an objective, independent existence?

It may be helpful to recall Korzybski and emphasize that whatever we say that order is, it isn't. It is more than we say, as well as being capable of being unfolded in infinitely many ways that are different. To attempt to attribute order solely to the object or to the subject is too limited. It is both and neither, and yet something beyond all this: a dynamic process that involves subject, object, and the cycle of perception-communication that unites and relates them. This approach suggests that no constitutive order is an absolute truth, for in reality its ability to lead to coherent and consistent activity is always limited.

While this may seem overgeneral and not a little abstract, a simple example will indicate its general trend. Some cities, such as New York, have regular grids of streets and avenues. In such cities the order of a grid fits harmoniously into the activity of walking through the city. But in a city of a more complex order, like London, such an imagined grid does not fit, and to continue its use, as a visitor from the United States may attempt to do, will lead only to confusion and frustration. In the case in which the grid pattern provides a satisfactory order for the activity of wandering through the city then it could be said to correspond to reality. But as this correspondence begins to fail, the walker will be alerted to the need for new acts of perception-communication and the creation of new orders. Clearly no one order will cover the whole of human experience, and as contexts change, orders must be constantly created and modified.

The example of the order of the grid was not chosen by chance, for in its form, as the Cartesian grid or coordinate system, it has dominated the basic order of physical reality for the past three hundred years. In many cases the Cartesian grid worked well, in the sense that it led to a coherent activity and thus corresponds to reality fairly well. However, in this book, the general appropriateness of the Cartesian description is called into question. Just as the New Yorker who travels to London will require a subtler notion of order than the rectangular grid of streets and avenues, so new orders are required to describe those aspects of reality that have revealed themselves during this century.

Chaos, Randomness, and Infinite Order

Let us return to the notion of order as similar differences between successive segments of a curve or other geometric figure. In order to expand this notion it is first necessary to introduce the idea of an *order of orders*, which leads naturally to the notion of a *degree of order*. In the previous examples, curves were described in terms of single differences, which could be made as complex as desired. Such curves are determined by two pieces of information: the location of the starting point and the common difference in successive line elements (this remains similar to itself throughout the curve). These curves therefore have an order of *second degree*.

Subtler curves, corresponding to orders of higher degree, can be defined when the differences themselves become different, but similar in a higher order. For example, consider the line below. The segments *a*, *b*, *c*, *d* all form an order in which the differences of successive segments are all similar. Similarly, the segments *e*, *f*, *g*, *h* form an order of similar differences. But the similarities that define these two successive orders are different, since the segments lie in different directions. The segments *i*, *j*, *k*, *l* also form an order. However, the difference between *a*, *b*, *c*, *d* and *e*, *f*, *g*, *h* is similar to the difference between *e*, *f*, *g*, *h* and *i*, *j*, *k*, *l*. In other words, there are two orders of similar differences underlying the curve below. In this fashion it becomes possible to generate higher levels of order which relate lower levels of order and in this way describe an *order of orders*.

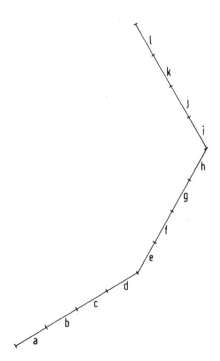

Since the above curve is defined by three items of information, the starting point of the first segment, the difference between adjoining segments, and the difference of the differences, it has an order of *third degree*. In principle such orders can be continued indefinitely to orders of higher and higher degree, and even to orders of infinite degree.

As pointed out earlier, the order of movement of a particle in Newtonian physics is of second degree. The motion of particles is normally described by a second order differential equation, derived from Newton's second law of motion. This means that the rate of change of velocity of a particle (that is, its acceleration) is determined once the nature of the external force is given. It is possible to analyze the movement into very small steps that succeed each other in short intervals of time. The actual change of velocity between one very small step and the next is therefore determined as being proportional to the force operating within that step. Indeed this statement is equivalent to Newton's second law of motion. In other words, if the force is defined throughout the particle's whole trajectory, then once the particle's initial position and velocity have been given, the entire motion and trajectory are determined.

In the case of a constant force, such as gravity or other forces commonly met with in nature, then the differences between successive velocities are equal. Hence the trajectory has an order of second degree. But what if the force happens to vary with position, even with time as well? What will be the order of the curve? The answer to this question opens up the whole discussion of what is the true nature of randomness and chaos in nature and it requires some careful working through.

In a way, even if the force varies with time and position, the order is still one of second degree. How can this be? The answer is that provided the actual law of force is known, then the trajectory is still determined by two items of information: the initial position and the initial velocity. For at each interval, knowing the law of force, the actual difference between segments can be computed. So that, knowing the initial values of position and momentum, a unique trajectory can be defined. However, in another sense, the motion is of much higher

degree, since the curve appears to be far more complex than
other second-degree curves. Indeed, if the value of the force is
not known for each segment, then a large number of parameters will be required to define the curve.

It appears that the motion of a body is capable of two
different descriptions, of being both of high and of second
degree at one and the same time. To see this requires the
examination of a more concrete example. Think of a ball that
rolls easily downhill. If the hill is smooth, like a flat plane
inclined to the horizontal at an angle, then if the ball is
released from rest it will roll down the hill in a straight line. If
it happens to be given a little push to the side, then the ball
will roll along a curved path. But in either case the motion has
an order of second degree. Now, suppose that the hill is highly
irregular, full of corrugations, bumps, rocks, saddle points,
and hollows. The trajectory of the ball will depend very sensitively on just where it is released and how it is set in motion. If
it is given even the slightest push to the side, on being
released, it may eventually come up against an obstacle which
will deflect it a great deal. A typical trajectory of the ball will
undulate and turn, undergoing a complex series of large and
small deflections in a very complex way. Clearly the similar
differences of velocity within successive small steps of the
motion are no longer constant. The ball meets an ever-changing
force, because of the irregularities of the hill, so that the
differences in velocity are constantly changing. Since the changes
in these velocity differences are themselves subject to change
in a complex way, it would appear that the order of the motion
must be very high indeed, certainly higher than the second
degree.

On the other hand, if all the details of the slope of the hill
are known, with every bump and hollow being defined, then
only two items of information are needed to define the trajectory. Once the initial position and velocity of the ball are
established, then a unique trajectory down the hill is established. A slightly different value to initial position and velocity may produce a very different trajectory; nevertheless only

these two items of information are required. In this sense the trajectory is also of second degree.

This apparent paradox can be easily resolved by proposing that the notion of order is in fact context-dependent. Therefore, in the context in which the details of the shape of the hill are not taken into account, the motion is of very high degree. But in the context in which all the fine details of the slope are known, the order is of second degree, because only two items of information are needed to define any trajectory. Clearly both kinds of order are relevant and both correspond, each in its own way, to the reality of the curve. This clearly indicates that order is neither subjective nor objective, for when a new context is revealed, then a different notion of order will appear. This example harks back to the three people walking through a forest. In each case the order of the forest they perceived depended upon the context of their overall activity.

In the limit in which the hill becomes excessively bumpy and irregular, the motion of the ball will appear to be that of infinite degree, or in more familiar language, random. Hence the above discussion suggests that there is a connection between randomness, chance, and chaos on the one hand, and order on the other. This can be seen in the case of the generation of random numbers. In order to carry out certain operations, computers sometimes need to call upon strings of random numbers and therefore they contain their own internal programs for generating them. A particularly simple program takes a given eight-digit number and multiplies it by itself. The resulting number will be very large but the program selects only the middle eight digits, which are then multiplied by themselves, the center digits taken, and so on. In this way a series of numbers are generated which do not appear to have any particular order to each other.

The program that generates these random numbers has an order of quite low degree. The numbers themselves are generated according to a determinate rule and in the context of this rule they could be said to be of low degree. However, in a context that does not include the computer program, the numbers appear to succeed each other in a very complex and unpredictable way. This sequence will be, as far as it is possible to test, free from all correlations and without any significant suborder. In this sense, therefore, the order of the numbers is essentially random. To put it another way, this set of numbers cannot be distinguished from any other set that is called random. Yet in the context of the computer program, a simple order of low degree determines the succession of numbers. Depending on the context, therefore, the order of these numbers is of infinite or of low degree. Again, it appears, the notions of a random order and an order of low degree depend upon the wider context in which they are embedded.

Recently this whole topic of chance and randomness has become the focus of a new mathematical development called chaos theory. In this theory a number of differential equations of the second degree have been shown to give rise to trajectories that possess orders of infinite degree. In addition to random curves, these trajectories also include subtler curves which are nonrandom but of very high degree. Such curves are extremely sensitive to initial conditions, like the ball rolling down the bumpy hill, and can wander across whole regions of space in highly irregular ways. They correspond to the intuitive notion of what constitutes chaotic motion.

An example of such motion is given by the ocean as it breaks on rocks near the seashore. At first sight this seems to be totally irregular, yet closer inspection shows many suborders of swirls, flows, and vortices. The word *chaotic* provides a good description for the order of such a movement. Within the context of order that is visible to the eye of a close observer, this motion contains a number of suborders and is far from random. Nevertheless, to a more distant viewer these suborders become so fine that they are no longer visible to the eye and the order would be called random.

More generally a random order can be defined as a special case of a chaotic order. It has the following characteristics:

1. It is of infinite degree.
2. It has no significant correlations or stretches of suborder of low degree.

3. It has a fairly constant average behavior and tends to
vary within limited domains. This domain remains more
or less constant, or else it changes slowly.

This definition of random order accounts well for the distribution of shots from a fixed gun. For to begin with, there is no
order of finite degree that can predict where an individual shot
will exactly hit the target. Second, the successive shots are not
significantly correlated. Third, the mean position, and mean
variation of this position, is fairly constant, since the shots will
scatter over an area that depends on the fine details of the gun,
ammunition, wind velocity, and other factors.

The shots from the gun are called random. However, if the
context is extended, then each shot becomes more nearly
predictable. For example, if the wind velocity is measured, or
if variations in the gun emplacement are observed, then more
information is available to determine this new context and
individual variations can be calculated. This emphasizes again
that the notion of randomness is inherently context-dependent.
This context can be either objective or subjective. For example, the variation of the wind velocity may be measured to
provide an objective context, or subjectively an observer may
define a new set of conditions. In general, however, the
context of order is a combination of both.

Randomness is being treated not as something incommensurable with order but as a special case of a more general notion
of order, in this case of orders of infinite degree. This may
appear to be a curious step to take, since chance and randomness are generally thought of as being equal to total *disorder*
(the absence of any order at all). This question of the meaning
of chance, randomness, and disorder has been a particular
headache, not only in science, but also in mathematics and
philosophy. But here it is proposed that whatever happens
must take place in *some* order so that the notion of a "total lack
of order" has no real meaning. Indeed even what are called
random events do happen to take place in a definable and
describable sequence and can be distinguished from other

random events. In this elementary sense they obviously have an order.

The notion of chance as a form of order can be illustrated by considering the random number generator in a computer. The actual sequence of these random numbers is generated by a deterministic sequence of instructions. However, the disadvantage of this procedure is that each time the program is activated it will generate an identical sequence of "random" numbers. If such a sequence were to be used in a video game, for example, then the chance movements of the rocketships would be the same in every game. One way of overcoming this obvious drawback is to begin the program each time at a different starting point, or to choose some starting configuration more or less by chance. For example, the setting on the electronic clock that monitors the computer's internal time could be used as a parameter in the program. Each particular time on the top would determine a fresh input parameter and, therefore, a new set of "random" numbers. As the clock setting changes, one random sequence is therefore replaced by another.

Each of these sequences of random numbers has a definite order of succession that can be distinguished from that of any other. In the context which includes the computer, its program, and the clock setting, each sequence is of an order of low degree. However, in the absence of such a context the sequences are of infinite degree and cannot be determined by any finite number of differences. It is clear that randomness cannot be equated with a complete absence of order, which in itself has no meaning. Rather, randomness is a particular kind of order which satisfies the three requirements that were given earlier.

THE FULL RICHNESS
OF THE SPECTRUM OF ORDER

Not all orders of infinite degree are random. More generally there is a broad spectrum of order, going from second to infinite degree which contains a very rich but little explored domain. Within this domain are to be found whole ranges of subtle and complex orders, some of infinite degree, which contain embedded within them many orders of lower degree. This hierarchical nesting of these suborders forms a greater order of its own.

Considerations of this kind make it possible to look in a new way at some traditional problems and questions concerning the whole notion of order. Language, for example, may be considered as having an infinite order, because its potential for meaning is unlimited and cannot be determined by any finite set of differences. On the other hand, it also contains many different suborders of lower degree—the various rules of syntax and semantics for example. The higher orders also contain and condition these suborders. Within the infinite order of language of a novel, for example, is contained the order of the sentence; the orders of tense, action, and the subject of the paragraph; and the orders of character and plot that link the chapters together. Each of these suborders, complex as they are, is not independent, for it is conditioned by the overall flow of the novel. Tenses may suddenly change, in a disjointed way, to suggest tension and rising action. Narration may change from third to first person to accompany a more subjective passage. Certain rules of syntax may be deliberately broken or distorted. Semantic meanings may be played with and even the basic sentence structure may be, at times, broken. All these various transformations within the suborders combine together to further the order of the novel itself.

The infinite order of language in the novel therefore contains a richness which is not predictable and cannot be fully pinned down within any finite series of differences and similarities. Although locally certain forms of order may be recognized, it is possible that at some other point the structure may change. Although language is of infinite order, it is clearly not random; rather it is intelligible and meaningful at a very high level.

An essential point in this discussion of the order of language is the context dependence of its meaning. Only within the context of a human being, with his or her capacities, knowledge, and experience can an order of meaning in what is read or heard emerge. If a particular person lacks this context, for example, if he or she does not know anything of the semantics of a particular language, then, at least initially, the order will appear to be nothing more than a rhythm of sounds. The overall order therefore belongs both to the language and to the person who uses it.

A similar dependence of meaning on a broad context is found in music which is thought of as being "modern" or "avant-garde." Such music may even be judged as "meaningless" or "offensive" by a listener who does not have the adequate context from which to perceive the whole order of the music and who attempts to understand its meaning in terms of an earlier and outmoded context. It is only as more and more of such music is heard that the listener learns about its different forms of structure, particular sonorities, and the composer's intentions. In this way a context is developed in which the music becomes meaningful and satisfying. Only now is the full order of the music unfolded by the listener.

The development of a context, in which a novel or a musical composition assumes its full meaning, depends strongly on the tacit infrastructure of ideas, knowledge, and skills that are available in a given community and subculture. Likewise, such an infrastructure plays a significant part in the case of scientific research as well. Obviously it is of key importance that such an infrastructure should not be maintained in a rigid and inflexible way. This, of course, holds as much for language and music, and indeed for every area of life, as it does for science. Without proper free play of the mind, the context provided by the tacit infrastructure will be far too limited for the creative perception of new orders.

The subtle orders of infinite degree discussed above are neither random nor simple regular orders. This implies that randomness can in fact be thought of as one aspect of a general spectrum of order. At one end of this spectrum are the simple orders of low degree. At the other are the random orders, and in between is a whole world of complex and subtle order, including language and music as well as other examples that could be drawn from art, architecture, games of all kinds, social structures, and rituals. But this discussion need not be limited to human activities alone. Clearly life itself is of such an infinite and subtle order. Such orders are also found in inanimate, physical systems. Consider, for example, the motion of a fluid, such as water, which is described by a differential equation. The motion is, in principle, determined by this equation, along with the initial movements of each element in the fluid and by the form of the boundaries of the fluid. Under the most simple conditions, in which the boundaries of the fluid, along with the fluid's initial movements, are smooth and regular, the fluid will flow in a correspondingly smooth and regular pattern which has an order of low degree. This can be seen in a slowly flowing river, which contains no rocks or obstructions. However, if irregular banks or obstructions are present, or if the water is initially agitated, then the same differential equation will predict a motion that may become chaotic. In an extreme limit, it might even become random.

The flowing river gives a good image of how a simple order of low degree can gradually change to a chaotic order of high degree and eventually to a random order. In this process, complex whirlpools may develop and the water may break up into foam, bubbles, and spray. The origin of this behavior lies in the relationships between the elements of the flowing water. Each element would, if left to itself, follow an order of low degree. But in fact, each is affected by all others, which are, for it, external influences that change its motion. These bring about a rapidly changing force on the element in question that leads to an order of high degree. In this sense the description of the motion is not unlike the example of the ball rolling down the hill. Here the element of fluid takes the place of the ball, and all other elements of the fluid play the part of the highly irregular slope and surface of the hill.

In this fashion it is possible to follow the change in order from the smooth flow of the river, through ever-increasing turbulence, to chaos and eventually into motion with no correlations or suborders that are significant within the context in question. Randomness is thus understood as the result of the action of the very small elements on each other, according to definite orders or laws in an overall context that is set by the boundaries and the initial agitation of the water. By treating randomness as a limiting case of order, it is possible to bring together the notions of strict determinism and chance (i.e., randomness) as processes that are opposite ends of the general spectrum of order.

In this connection, it is important to emphasize that although the order of a random sequence is of infinite degree, it does not have the subtlety possessed by the infinite orders of music, art, and language. A typical random order is generally of a relatively mechanical nature. It can as a rule be correlated to mechanical causes that are to be found in a broader context. This was discussed in the case of the distribution of gunshots and of the random numbers generated by a computer program. A similar but more complex case is that of Brownian motion. This is the motion in a random path that can be seen in very small particles, such as pollen grains, immersed in water. By itself, the natural motion of a pollen grain is an order of second degree. However, because this particle is acted on by repeated molecular collisions, the effect is to disturb this second-degree straight line, and to transform it into an order of infinite degree. This is the result of the action of forces that are external to the particle which are very complex and multiple in nature, namely, the impact of countless molecules.

As in the case of a ball rolling down a hill, chaotic motions arise from the action of determinate forces. This conclusion is reinforced in the case of systems of many particles. Each particle is subject to forces arising from the others that are, for it, contingencies that vary in an almost infinitely complex way. In a context in which all these forces are taken into account in detail, it is possible in principle to obtain a deterministic account of the inner movement within the system. In such a case, the forces acting on each particle are specified and so its trajectory can, in this context, be regarded as an order of second degree. In a context in which all these details are not taken into account, the order will be of infinite degree. It can also, under suitable conditions, satisfy the three criteria for randomness that were given earlier. This can in fact be demonstrated mathematically for certain simple but typical kinds of interactions between particles.

A random order can thus be explained within such a system without the need to take into account any contingencies which are external to the whole system. From this standpoint randomness or chance is *necessary*, and this necessity is not subject to external contingencies but is an *inner* necessity. This leads us to propose a further metaphor: Chance *is* necessity (of a particular form).

The above treatment, while valid up to a certain point, is, however, still an *abstraction* and an *approximation*. For no system can correctly be regarded as totally isolated and self-determined. Thus, most systems of appreciable complexity are capable of developing instabilities, so that they may be profoundly affected even by weak external interactions. And even more important, no specified statement of the laws of nature will be completely and universally valid. For, as has been pointed out many times, whatever we say anything is, it isn't—it is also something more and something different. A more comprehensive law would leave room for this by allowing even the most basic orders known to be at least potentially of very high or infinite degree. In extreme cases, these would approach chaotic or random orders.

In the limit of large numbers, such random orders can approximate simple causally determinate orders of low degree. Insurance statistics are a simple example of this, and the deduction of the laws of thermodynamics from a statistical treatment of the mechanics of molecules is another. It is in this way possible to reverse the original metaphor and to say (at least in certain contexts) that necessity *is* a particular form of chance.

This implies the interweaving of simple orders of necessity and infinite orders of randomness in a potentially infinitely complex structure of law.[1] What is randomness in one context may reveal itself as simple orders of necessity in another broader context. And vice versa, what is a simple order of necessity in one context may reveal itself as chance in another broader context. But in a still broader context, both are to be seen as extremes in the rich spectrum of orders of varying degrees that lies between them. Thus, there is no need to fall into the assumption of complete determinism (although this may in certain fairly broad contexts be a correct abstraction and approximation). Nor is there any need to assume that chance and indeterminism rule absolutely (though these too will provide correct abstractions and approximations in their appropriate contexts). No matter which system of law may be appropriate in the context that is currently under investigation, there is always room for something more and something different— something that will be more subtle and that has the ultimate potential for being a manifestation of creativity.

THE QUANTUM THEORY AND "HIDDEN" ORDERS

In the quantum theory (at least as this is usually interpreted), it does not seem to be possible to maintain the notion of the interweaving of the orders of chance and necessity as abstractions from infinite order with unlimited creative potential. The principal difficulty arises because a system of particles cannot simply be treated by analysis into independently existent but interacting constituent particles. Rather, the quantum theory implies a certain quality of wholeness in the sense that a system cannot be dealt with properly as a set of separate parts. Hence in the usually accepted interpretation, there is no way to discuss how randomness might arise. Randomness is therefore assumed to be a fundamental but inexplicable and unanalyzable feature of nature, and indeed ultimately of all existence. Such an approach complements Bohr's notion of the inherent ambiguity of concepts at the quantum mechanical level, which was discussed in Chapter 2. For within the range of this ambiguity, the quantum theory implies that the experimental results will fluctuate with an irreducible kind of randomness. And thus the very nature of quantum mechanical ambiguity will imply a corresponding limit to the possibility of meaningful order.

1. For a discussion of this, see David Bohm, *Causality and Chance in Modern Physics*, Routledge and Kegan Paul, London, 1957.

This book, however, proposes the notion that everything that happens takes place in some order (which, however, depends on broader contexts for its meaning). Therefore, while there is ambiguity within particular contexts, the notion of an ultimate limit to the meaning of order that holds in all possible contexts is not admitted. For example, in the previous chapter the causal interpretation of the quantum theory was discussed in which a further kind of order is proposed. This order, which underlies the randomness implied in the probabilistic laws of the quantum theory, can be understood as the causally determined motion of the particle under the quantum potential. Because this latter potential gives rise to a complex and highly irregular force, the motion will in general be fairly chaotic. In such complex systems, an essentially random order is to be expected which explains the probabilistic assumptions of the theory.

However, as proposed at the end of the discussion on the causal interpretation of the quantum theory in the previous chapter, such a simple deterministic theory is an abstraction, which is valid only up to a point. Beyond this point, one may have to consider the possibility that even the basic laws of the particles involve orders of infinite degree, which reflect levels of reality beyond those treated adequately by the current quantum theory. In a certain approximation, this may be considered as a random order. But as we have indeed already suggested earlier, the motion may more generally be in an order of very large or infinite degree, which is in the rich domain between simple orders of low degree and infinite chaotic or random orders. When understood in terms of the causal interpretation, the quantum theory is therefore capable of fitting into the general notion of the interweaving of chance and necessity, considered as lying at the extremes of an infinitely rich order that is context-dependent. In quantum theory this order is at present hidden in the contexts available so far in physics, because science has not been able to respond with the necessary subtlety of meaning. But in contexts that may one day be discovered, new possibilities for creativity within such orders may be revealed, and these context orders will then cease to be "hidden."

The notion that both simple regularity and randomness in a given context may contain orders that are "hidden" in that context, but which can be revealed in other contexts, is of quite general significance. For example, the order of a language or music cannot be found by studying the regular orders of vibration in sound waves, or the almost random orders of motion of the molecules in the air that carries them. Indeed, unless the mind is free of rigid commitments to familiar notions of order, such as the kind described above, it cannot provide a context within which basically new orders might be perceived. When music and art explore new orders, these are not even apprehended by those who are rigidly habituated to the older and more familiar ones. It is quite possible that in nature, there are further new orders, going beyond those that can be comprehended in terms of the quantum theory, even with its causal interpretation, as extended by the notion of orders of infinite degree in the motions of the particles.

Some examples of these will be explored in the next two chapters, notably the generative order and implicate and superimplicate orders which may be relevant for the understanding of life, consciousness, and intelligent perception. Nevertheless, there is little or no room for such orders within currently accepted notions of physics, chemistry, biology, and other sciences. So, in terms of present conceptions, whatever could be the basis for such order in natural processes would probably be apprehended as "no order at all"; that is, what is commonly called disorder or randomness. It should therefore be clear how important it is to be open to fundamentally new notions of general order, if science is not to be blind to the very important but complex and subtle orders that escape the coarse mesh of the "net" of current ways of thinking.

ORDER FROM CHAOS AND CHAOS FROM ORDER—
THE MEANING OF ENTROPY

In addition to the transformation of order into randomness, as
discussed above, there is a corresponding transformation of
randomness into order, which has been much studied by Ilya
Prigogine. Prigogine considers systems that undergo random
molecular motions away from equilibrium. Within such systems, a gradient or flow is established. For example, a difference in temperature may result in a flow of heat, or a difference
in chemical potential may result in a flow of atomic or molecular ions. Such a flow corresponds to an order of very low
degree, generally the second. Given the conditions of an order
of infinite degree—the random molecular motions—on which
is superimposed an order of very low degree, there appears a
very complex but regular array of patterns which undergo
systematic movements. In the case of a temperature gradient in
a liquid, this can take the form of what is known as the Bénard
instability, a spectacular phenomenon in which, in the words
of Ilya Prigogine and Isabelle Stengers, "Millions of molecules
move coherently, forming hexagonal convection cells of characteristic size." In the case of chemical gradients, a number of
complex oscillating reactions, such as the Belousov-Shabotinsky
reaction, are produced. Many other examples of the emergence
of global orders out of underlying chaos are discussed in
Prigogine and Stengers's book.[2]

Such transformations between randomness and simple regular orders are intimately related to the entropy of a system. The
notion of entropy is a concept of particular importance, not
only in physics but in chemistry and the life sciences. Entropy
is popularly described as the measure of disorder in a system,
a notion that clearly carries subjective overtones. On the other
hand, the science of thermodynamics enables the quantity
known as entropy to be measured objectively in terms of the
amount of heat and work that is associated with a system. Left
to itself, a physical system tends to maximize its entropy, a
process which is therefore associated with decay, disintegrat-

2. *Order Out of Chaos*, Bantam, New York, 1984.

ing, "running down," and increasing disorder in the system. But according to the metaphor that chaos *is* order, an increase in entropy has to be understood in a different way, that is, in terms of a kind of change of order.

Of key importance in this connection is the idea of a *range of variation* in random and chaotic motion. This idea was introduced earlier in the case of the grouping of shots from a gun. A more interesting example, however, arises from a river that is in chaotic movement. Imagine an irregular and changing whirlpool that fluctuates in a very complex way, but always remains within a certain region of the river. The whirlpool may perhaps be roughly determined by neighboring rocks or features in the riverbed. As the velocity of the river increases, this variation in space may grow. But in addition, there will also be an inward growth of subvortices of ever finer nature. Therefore, a measure of the overall range of variation of the whirlpool should include both of these factors—the inward and the outward growth.

As a matter of fact, in classical mechanics, a natural measure of this kind has already been worked out. Its technical name is *phase space* and its measure is determined by multiplying the range of variation of position and the range of variation in momentum. The former, the range of variation in position, corresponds roughly to the changes in location of the vortex as it spreads out into the river and the surrounding water becomes more agitated. The latter, the range of variation of momentum, corresponds to the extent to which the whirlpool is excited internally so that it breaks into finer and finer vortices.

Clearly the measure in phase space corresponds quite well to an intuitive notion of the overall degree of order involved in the flow. For the more the general location of the vortex expands, the higher is the degree of order; and the more the internal vortices subdivide, the higher also is the degree of order. What is particularly interesting about this measure in phase space is that it corresponds to what is actually used in physics to define entropy.

Entropy is a concept which is of vital importance in many areas of science, yet which lacks a clear physical interpreta-

tion. For example, there has been much debate on the extent
to which the concept of entropy is subjective or objective.
However, with the present approach to the notion of order,
chaos, and randomness, it is now possible to clarify what is
meant by entropy.

Consider an isolated system of interacting particles. Each
particle acts as a contingency for all the others, in such a way
that the overall motion tends to be chaotic. When such a
system is left to itself, it moves toward what is called thermal
equilibrium, a condition in which there is no net flow of heat
or energy within the system and regular suborders vanish
almost entirely. In this state of equilibrium, the entropy of the
system is at its maximum. This maximum entropy is therefore
associated with the inability of the system to carry out work,
transfer useful energy from one region to the other, or in any
other way generate global orders of activity.

In statistical mechanics the numerical value of this entropy
is calculated from the range of random motion in phase space.
(To be more exact, it is the logarithm of this measure.) This
means that when energy is added to the system, the range of
random motion will grow and the corresponding entropy will
increase.

A change in entropy is therefore a measure of the change in
the range of fluctuations that occur within the random order.
Entropy now has a clear meaning that is independent of subjective knowledge or judgment about the details of the fluctua-
tion. This approach to entropy does not require any discussion
of disorder, which in any case cannot be defined in a clear
way. Treating entropy in this fashion avoids many of the
difficulties normally associated with this topic, such as the
subjective notion of what could be meant by disorder. After
all, since entropy is an objective property of a system which
can actually be observed with the aid of thermodynamic processes, why then should subjective and ultimately undefinable
feelings about disorder affect the objective behavior of such a
system?

To sum up, the question of randomness is an aspect of the
general context dependence of order. In a microscopic context

that takes the details of the forces between the particles into account, a particular thermodynamic system may have a well-defined order of its inner movements, of quite low degree. Nevertheless, in a macroscopic context that does not take such details into account, the same system will have an order of infinite degree in its random fluctuations. These determine its entropy and therefore its general thermodynamic properties. Both orders are equally objective and equally subjective.

In this connection, it would be clear that this discussion embraces many of Prigogine's notions. Prigogine considers his basic idea to be the emergence of order out of chaos. Here, this is described as the emergence of orders of low degree out of a chaotic order of infinite degree. Indeed what Prigogine calls "chaos" is not actually *complete* chaos, but rather, it is an initial random order on which is superposed yet another initial order of low degree. Out of this complex interweaving of the original infinite chaotic order with the order of low degree emerges yet another order of low degree. Thus, the whole process is basically a transformation between one overall order and another (in which the net entropy is increased, in spite of the emergence of the new order of low degree).

More generally in physical systems there is a whole spectrum, with orders of low degree at one end and chaos and randomness at the other. In between are further kinds of order of great subtlety that are neither of low degree nor chaotic.

Science, however, has not yet explored these intermediate orders to any significant extent. They may turn out to be quite important in many areas and indeed life itself may depend on them.

Until now, science generally has regarded orders of low degree and random orders, as being "incommensurable" or mutually irrelevant. This leads to breaks in communication and continuity between those aspects of research which use these orders in different ways. There is already, however, a kind of connection in which causal orders are treated as emerging from random orders in the limit of large numbers. Insurance statistics are a simple example of this, and the deduction of laws of thermodynamics from a statistical treatment of the laws of the mechanics of molecules is another. More recently, with the emergence of chaos theory, it has become clear that it is possible to go in the other direction, and treat statistical laws as emerging from causal laws. However, it is being proposed in this book to start from the whole spectrum of order, and to consider causal laws and statistical laws as being limiting cases. In this way there is no break in communication, and fields which concern themselves with different parts of this spectrum will now have a common conceptual basis so that creative communication is in principle possible between them.

To return to the question of social order discussed at the beginning of this chapter: It is now possible to explore the question of whether the eternal order and the secular order can be similarly regarded as two extremes of a spectrum, between which lies a rich field in which new orders of society could be creatively perceived. More generally, an approach carried out in this spirit could perhaps embrace different social orders that at present cannot meet, and might help to avoid the irreconcilable conflicts that are now arising between such orders.

STRUCTURE

The concept of order is, by itself, of very general interest. But one of its most fundamental and deepest meanings is that it lies at the root of structure, which is a key issue, not only in science, but in life as a whole. Structure is often treated as being static and more or less complete in itself. But a much deeper question is that of how this structure originates and grows, how it is sustained, and how it finally dissolves. Structure is basically dynamic, and should perhaps better be called *structuring*, while relatively stable products of this process are *structures*. But even these latter structures should not be considered as basically static, for they are the results of processes which sustain them and keep them, for a time, more or less within certain limits.

As with order, so with structure there can be no complete definition. Rather, to put it again: Whatever we say structure is, it isn't. There is always something more than what we say and something different. At any given stage, it is possible to abstract a certain structure as relevant and appropriate. But later, as the context is made broader, the limits of validity of this abstraction are seen and new notions developed. In the time of the ancient Greeks, matter was commonly abstracted as having a continuous structure, but later there arose the abstraction of a discrete particle structure. In the nineteenth century this, too, was seen to be limited, and deeper continuous field structures were proposed. With the advent of quantum mechanics arose a further abstraction of structure which went beyond the dichotomy of the continuous and the discrete. In the future, as the context is extended even further, still newer notions of structure may arise in a similar way.

Structure itself is based on order, but involves much more. According to the dictionary, structure is the order, arrangement, connection, and organization of simpler elements. But it must be emphasized that these "elements" are not necessarily separate physical entities. More generally, they are terms introduced in thought for the sake of conceptual analysis, as with the elements of fluid in the river that were discussed in an earlier example.

For the sake of illustration, structure will first be developed in terms of simplified elements that have a separate existence. But it must always be remembered that, at a deeper level, attention must be given to the whole, which, in turn, acts to guide thought as it abstracts elements which do not in fact have a separate existence. Consider the example of the novel discussed in the previous section. While its use of language illustrates a complex and subtle infinite order, it is more comprehensively described as being a structure, but one of infinite complexity. The various suborders within the novel, of tense, mood, character, location, and so on, are all arranged, connected, and organized together. Yet each suborder, or element, is clearly inseparable from the greater whole. In a similar way there could be said to be structure in music or in a painting.

This method of conceptual analysis of structure makes it

possible to unfold the whole notion, to articulate it, and to connect it to the notion of higher order. To begin, it is possible to go from a simple linear order to an *arrangement* of such orders. This involves putting similar orders together. The system of parallel lines below is clearly such an arrangement. Each line is characterized by a set of similar differences and its relationship to other lines gives a further set of differences that are similar. The arrangement of lines is, therefore, an order of orders. Such a notion could be developed further to give a hierarchy of orders, which clearly would form an important component of structure.

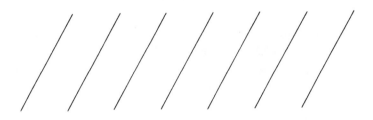

But the above arrangement of parallel lines could by no means be called a structure. What is needed is some *connection* of the elements. For example, bricks in a wall are arranged in an order and in an order of orders, but they are also in contact so that they make a wall. In turn, walls in contact make a room and rooms in contact make a house. In many such examples, contact is of a local nature in which neighboring elements touch. However, structures can also be created through nonlocal contact in which elements may be physically separate but held together by, for example, electrical or magnetic fields. It is therefore possible to arrive at the notion of a static structure which involves a hierarchy of orders, together with some form of local or nonlocal contact.

But to return to the more basic dynamical view of structure: Even in the case of something as static as a house, it is necessary to ask how it comes to be built, how it is maintained, and how it eventually falls and is destroyed. It is clear then that any such structure is subject to a process of *organization* and *disorganization*. This includes, in the first instance, the overall principles (as supplied, for example, by an architect) which determine how the suborders are to be organized to fit together, with appropriate arrangements and connections. To these must be added the processes that are involved in actually bringing about, sustaining, and breaking down such an organization of structure.

The house is not a particularly graphic illustration of this key dynamical aspect of structure. A better example is to be found by considering life and intelligence. Thus, in a tree, a structure is clearly visible. For not only are there the many interrelated orders and arrangements of trunk, branches, twigs, and leaves, which we have discussed earlier, but these are also connected onto a single whole. This whole is organized through the processes of metabolism, in which the tree is formed and sustained, and eventually dies. According to current theories, the DNA molecules carry what is equivalent both to the architect's plans and to instructions needed for growth, maintenance, and repair. With living animals, this process of organization is much more complex and it depends on several systems, including a brain and a nervous system. Finally, with intelligent beings, new levels of organization appear, involving many very subtle kinds of structures, such as language, music, thought, and so on, which all contain orders, arrangement, and connections of elements organized in an extremely dynamic way.

It must be emphasized again that the stability of structure is not static but arises through a form of mobility in which any forces which tend to break the structure down are compensated by processes taking place within the structure itself. It is evident that this notion of stability of structure through mobility is of crucial importance not only for an understanding of inanimate matter but also for living beings, consciousness, and society.

RATIO OR REASON

The understanding of structure and its expression in thought and language takes place primarily through *reason*. The word *reason* is based on the Latin *ratio*. A little consideration shows that a kind of ratio is indeed a key feature of reason. The general form of ratio may be written as *A:B* as *D:C*, with the numerical ratio $A/B = D/C$ being a special form of this. Such a ratio means that *A* is related to *B* as *C* is related to *D*. However, two things can be related only if they are different. But in Latin, the root meaning of *difference* is "carrying apart." To "relate" comes from the past participle of "to refer," which means to "carry back." In this process two things are, at least in the mind, carried apart to difference and then carried back to similarity and relationship.

The order of the line that was introduced earlier can therefore be described by the ratio *A:B* as *B:C* as *C:D*, and so on. Further, since any ratio can be represented by the symbol *R*, it is possible to relate ratios in a similar way, $R_1:R_2$ as $R_2:R_3$ and so on. Hence from a simple ratio, a relationship or relationships can be defined.

The full development of such a hierarchy of ratios or relationships, which occurs in all areas in which the mind is used, is essentially the power of rational thought or reason. Irrationality can then be taken as the failure of such ratios to cohere. Rationality is thus an order, and indeed is the essential order of thought.

Ordinarily a test for rationality is made with the aid of logic (which is generally taken to be a set of formal rules that thought must satisfy if it is to be judged rational). The common attitude to such formal logic is to regard it as a static set of norms, which stands entirely on its own. Moreover, formal logic is in fact only a limiting aspect of a much broader, overall movement of reason. In harmony with the general approach to science in earlier chapters, it can be said that in its origin, the broader movement of reason is basically a *perceptive* act and that formal logic is a relatively fixed abstraction from this movement. The German language illustrates this better than English does, because its word for reason is *vernuntt*, based on the verb *vernehnen*, which means to perceive, with the connotation of perceiving through the mind. This can be rendered into English as "intuitive reason" or "perceptive reason." The corresponding word which, in some sense, covers formal logic would be *verstand*, meaning in this context not "to understand," in the sense of comprehension, but rather "to stand firm."

Clearly it is necessary that thought should have the possibility of standing firm, if it is to function properly. But this "firm standing" must find its appropriate place in the broader context of the flowing movement of intuitive reason. It is only in this broader context that thought can become the vehicle of creative perception. Indeed, when there is free play of the mind, thought has its ultimate origin in such perception. It then unfolds in a natural way, through proposition, composition, supposition, and disposition. It is transformed into something fairly well defined and, as it were, crystallized. Such crystallization of reason, which is ruled by formal logic, is indeed absolutely indispensable if the proposals in which thought begins are to be tested properly for their rationality and for their coherence with the actual fact. Nevertheless, formal logic has to be ready to dissolve back into flowing reason, whenever a sustained contradiction or opposition develops in the application of its relatively fixed forms. In such a case the mind will be able to respond with creative intelligence, to perceive new orders and new categories that generally lie "between" the static and unrelated extremes presented by pure logic (for example, simple orders of low degree and chaotic orders of infinite degree).

The above is a description of the proper order of rational thought. If, however, the mind is rigidly attached to fixed categories and orders, then, as has already been seen through many examples, the free play needed for such a rational order becomes impossible. Instead the mind is caught up in false play, from which a creative response may be impossible.

However, it should be added that, as suggested in Chapter 1, the false play can take many subtle forms that are difficult to detect. One form is to see truth as absolutely identified with formal logic. But another is, similarly, to identify truth with intuition and to fail to be open to the need for rigorous tests of this intuition, under appropriate circumstances, with the aid of formal logic. Moreover, rigidly fixed opinions, which are actually based on the misuse of formal logic, often present themselves in a somewhat vague and undefined guise that pretends to be genuine intuitive perception. This is especially common in the case of prejudices, that is, pre-judgments. They are evidently of a basically intellectual character but are nevertheless generally experienced in a deceptive way as intuitive perceptions and feelings. It follows then that the proper function of reason requires a creative intelligence that is free of every kind of excessive fixing of thought, in whatever form this may appear.

Mathematics is an especially significant example of the interweaving of intuitive reason and formal logic in the kind of process that has been described above. In this connection, it is interesting to note that the mathematician von Neumann defined mathematics as the relationship of relationships. Evidently this implies an indefinitely extended structure of thought, which is in some ways similar to a hierarchy. This structure is formed in a process in which relationships of one kind are interwoven with relationships of other kinds, while this whole is organized by relationships of yet different kinds, and so on without limit. The really creative act of a mathematician is to perceive the germ of this vast structure of relationships, and to unfold it into an ever more developed structure of thought that is constantly tested for coherence against the rules of formal logic.

It is clear from the above discussion that ratio or reason is the essence of mathematical structure. But such ratio can be discovered in all areas of life. Thus, a person can find a complex structure of ratio in his or her experience of nature: for example, in the flowering movement of perception of the ever-changing orders to be observed in the wood, as was discussed earlier. Similarly, there are such structures of ratio in a house, a crystal, the panorama observed from a high mountain, a human body, a painting, the use of language, and society itself. Such ratio is grasped intuitively as well as through the intellect. Its field is not exhausted through sense perception alone, for it also applies to the inward perception of feelings. Thus, a person may say of an emotion that it is, or is not, in proportion to the situation that generates it. Indeed whatever we apprehend is apprehended through some form of ratio. For example, to recognize anything whatsoever is to see that as various ratios are related in the object, so are they related to our mental concept of it. This is of course just what is also done in mathematics and in its applications.

Mathematics, however, has the advantage of being able to discuss pure ratio (for example, ratios of ratio) without requiring a specified substratum lying in some object or sensory experience. In some areas this ratio may be so clearly defined that it permits an unfoldment of long chains of inference, whereas in other areas these chains are relatively short. But, as indicated in the Introduction, this is done at the expense of going to high levels of abstraction.

A key form of ratio is analogy, which is, according to the dictionary, a kind of proportion. As thought develops from the initial flash of creative perception, the ratio in it unfolds and articulates and so develops into a structure, in the way that was earlier described. This kind of structure is essentially an analogy to whatever the thought is about.

If the analogy is good, the "proportions" in the ratios of the thought are similar to those within the object of thought; otherwise it is a poor analogy. This leads to the suggestion that thought moves naturally toward the improvement of analogy.

The notion proposed above brings us back once again to Korzybski's statement that whatever we say anything is, it isn't. For after all, no analogy is equivalent to the object itself. Every analogy is limited. And if what we say is an analogy, then the object cannot be what we say. However, the essential proportions or ratios in both may be similar, but there is always room for newer and better analogies.

To test the success of such analogies involves the cycle of activity that was discussed in earlier chapters. Each thought,

WHAT IS ORDER? **149**

with its content of analogy, gives rise to a disposition to act, which contains within it a set of proportions or ratios similar to those in the thought. The action is therefore imbued with a similar ratio or reason. The fact that objects so generally respond in harmony to such action suggests that everything must likewise contain something at least basically similar to ratio, and that this is what makes it possible for the world to be intelligible to human beings. Thus, intelligence, which includes creative perception of ratio, and intelligibility are not two separate and unrelated qualities, but rather are inseparable aspects of a single overall reality.

The fact that there is such an intimate relationship between human intelligence and the intelligibility of the universe can be understood in terms of a notion, commonly held during the Middle Ages, that each person is a microcosm, and thus stands as an analogy to the whole cosmos. This would explain how such a person could, through intelligent perception of ratio, produce analogies to whatever exists in the universe and even to the universe itself. For if this person already *is* an analogy to all this, then looking outward and looking inward will be two sides of one cycle of activity in which any aspect of the totality can in principle be revealed.

SUMMARY

The basic theme of this chapter is the proposal that order pervades all aspects of life and that it may be comprehended as similar differences and different similarities. An essential distinction was introduced between constitutive order and descriptive order, while at the same time it was noted that any actual order lies in a kind of spectrum between these limits. Order is therefore neither solely in the subject nor solely in the object, but instead in the cycle of activity that includes both.

Orders of varying degrees were then explored, leading to those of infinite degree and including all sorts of very subtle orders, such as those in language and in music. Order in general was seen to lie in a spectrum between simple orders of low degree and chaotic orders of infinite degree of which randomness is a limiting case. Indeed there is no place in all this for the concept of disorder but only for random orders of infinite degree that are free from significant correlations and suborders of low degree. In this fashion, it is possible to discuss not only the emergence of orders of low degree out of chaos, as treated by Prigogine, but also the inverse process of the transformation of orders of low degree into chaos. This enables entropy to be considered as a particular feature of the general order of movement.

Structure was treated as an inherently dynamic notion, which includes not only the order of whatever elements are abstracted in thought, but also an arrangement, connection, and organization of these elements. Each structure was considered to be stabilized as the result of the mobility of whatever are regarded as its elements. The chapter concluded with a discussion of how structure is comprehended through a hierarchy of ratio, which may be apprehended in a perceptive act of intuitive reason.

FOUR

THE GENERATIVE ORDER AND THE IMPLICATE ORDER

Up to now, order has been considered as arising, basically, through a sequence of successions. This is indeed a very common form of order and perhaps the one that is most familiar. In this chapter, however, another kind of order, called the *generative order*, is introduced. This order is primarily concerned not with the outward side of development, and evolution in a sequence of successions, but with a deeper and more inward order out of which the manifest form of things can emerge creatively. Indeed this order is fundamentally relevant both in nature and in consciousness. In the following chapters its relevance to society will also be discussed.

The generative order will be explored with the help of a number of examples drawn from mathematics, physics, and the fields of art and literature. This will lead, in turn, to the *implicate order*, which is a particular kind of generative order that has been most fully worked out in physics. However, the implicate order will be found to have a broader significance, not only in physics but also in biology, consciousness, and the overall order of society and each human being.

Although specific proposals for how the generative order may be used will be discussed, it is not the main purpose, in introducing this new notion of order, simply to pursue its application in detail. Rather, it is to use these ideas in order to go more deeply into the meaning of creativity. In succeeding chapters, these notions will provide a base from which to move yet further in the general direction of creativity.

FRACTAL ORDER

In the previous chapter, order, as discussed in terms of similar differences and different similarities, was considered largely as a means of understanding curves, structures, and processes that are already present in nature or in the mind. However, it is equally possible to use such a notion of order, based on similarities and differences, to *generate* shapes, figures, forms, and processes. For example, starting from a single segment it is possible to generate a line by means of a process of repetition, in which each element is similar (equal to) the next. A polygon can be produced through a similarity of angle and length. In a related fashion all second-degree curves can be generated from an initial difference which is repeated in a way that is similar to itself. Higher-degree curves require the repetition of more differences, but they can all be constructed in the same fashion.

This idea could be pursued in ever greater refinement. However, for the purposes of this section, a more developed form of order will be used: the mathematical theory of fractals, which was recently invented by B. B. Mandelbrot,[1] which is closely related to the theory of chaos, as discussed in the previous chapter. Fractals involve an order of similar differences which include changes of scale as well as other possible changes. A simple example is to start with a base figure, the triangle:

and then consider a generator, which is really a small triangle that can be applied to each side of the basic figure.

1. *The Fractal Geometry of Nature*, Freeman, New York, 1983.

In this way a six-pointed star is produced:

In the following step, the generator is reduced in scale and applied again to each line segment, giving rise to the figure:

and then to

Clearly this process can continue indefinitely and results in a figure with extremely interesting properties. The reader may turn to Mandelbrot's book for details but, for the moment, accept that the circumference of this figure has grown to be infinite and has no slope.[2] These are particularly curious properties to have been generated in such a straightforward fashion.

By choosing different base figures and generators, but each time applying the generator on a smaller and smaller scale, Mandelbrot is able to produce a great variety of shapes and

2. For the more mathematically minded, this figure possesses no first derivative.

figures that have very interesting mathematical properties. Some of these have the appearance of islands, mountains, clouds, dust, trees, river deltas, and the noise generated in an electronic circuit. All are filled with infinitesimal detail and are evocative of the types of complexity found in natural forms. In addition, they reflect the way in which the details of a form appear to be similar over a wide range of scales of size: Often when we "zoom in" on some object in nature it continues to exhibit similarities of form at greater and greater magnification. Other fractals show ever new detail at smaller and smaller scales.

Mandelbrot points out that the geometry of fractals lies much closer to the forms of nature than do the circles, triangles, and rectangles of Greek geometry. It could be said that traditional geometry, out of which much of mathematics and the tools of physics have evolved, is, in fact, a highly artificial way of describing the world. Something closer to the fractal order, on the other hand, should be an appropriate starting point for discussing nature in a more general way, and for providing better formal descriptions of the processes of physics and biology.

The complex figure generated from the triangle is a little like a very irregular island which, of course, possesses a coast line that is ultimately infinite in length, when analyzed on an indefinitely fine scale. Other fractals begin as simple lines which expand in highly subtle ways until they appear to cover the entire page. An interesting question is therefore generated by these fractal figures: What is their dimension? Are they lines, of one dimension, or planes, of two dimensions? The answer is that a fractal is of *fractional dimension,* lying somewhere between a line and a plane. (Other fractals may have a dimension that lies between that of a point [zero] and a line [one].) Indeed Mandelbrot argues that the *fractal dimension* of an object is a significant characteristic and, for example, a river delta or a country's coastline can be characterized by its particular fractal dimension.

But how can a geometrical figure, drawn on a piece of paper, have a fractional dimension? Consider a plane, this page for example. If a dot *A* is made on this plane, then any neighboring point *B, C, D,* or *E,* no matter where it is printed on the page, will also be in the plane.

                    •B

             •D •A   •C

      •E

This is not, however, true of a simple line *XY*. Although a point *A*, for example, is on the line, and the neighboring points *B* and *C* are on the line, it is always possible to find neighboring points *D, E,* and *F* that are not on this line. Hence one property of a line, which has one dimension, is that points in its immediate neighborhood can be found that do not lie on it.

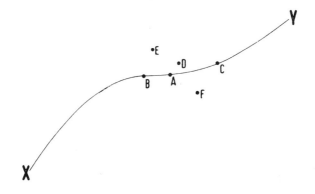

Now consider a fractal line with its unlimited complexity. As the fractal generator is successively applied, more and more points that previously lay outside this line will be included. Clearly, in some sense, it has more than one dimension. In the limit, in which the fractal line fills the plane so that no point remains in the plane that does not also lie on the fractal line, it will have become two-dimensional. So, in general, the dimensions of a fractal line lie somewhere between one and two.[3] And in three dimensions, general fractals can be constructed whose fractional dimension lies between zero and three.

While the fractal figures illustrated so far appear quite complex, they could hardly be called disordered, for they are composed of a quite simple order involving a single similar difference that is repeated at constantly decreasing scale. Moreover, figures of even greater complexity can be created using more than one generator and applying the alternative generators according to some fixed rule. One such rule of application, selected by Mandelbrot, is to use random numbers generated in a computer. In this way, through the introduction of random successive differences, he is able to generate the curves for Brownian motion as well as totally irregular coastlines.

It should be possible to generalize Mandelbrot's ideas still further by introducing additional categories of differences other than simple scaling, for example, differences in direction, shape, and so on, to arrive at yet more subtle fractal figures. Indeed, the principles involved in producing fractals may be much closer to those employed by nature than those associated with the figures and structures of traditional geometry. However, since so much attention has been given in the past to sequential order, it may be some time before a large number of concrete applications of Mandelbrot's ideas are discovered. Rather, the overall notion of generative order should be regarded as a very fruitful area for investigation, which may reflect not only on science but on many aspects of life.

---

3. Actually fractals drawn on a page will generally have between zero and two dimensions, for it is possible to generate fractals that have a lower dimension than that of the line.

GENERATIVE ORDER

Mandelbrot's fractals are only one example of a generative order (in the fractal case, a generation which proceeds by repeated applications of a similar shape but on a decreasing scale). Many other generative orders could be constructed in mathematics. However, the whole idea of generative order is not restricted simply to mathematics but is of potential relevance to all areas of experience.

Generative order can, for example, be seen in the work of a painter. Indeed, in a certain restricted sense the generation of form using Mandelbrot's fractals can be compared with the various stages of painting. At least until this century an artist did not generally begin to work with detail but, in the case of a portrait for example, attempted to capture the overall form and gesture of the sitter with an initial sketch on the canvas. Such a painter may have even employed the trick of squinting at the sitter in order to cut down detail and emphasize tone and shadow. Gradually this initial sketch was built up and made more detailed, solidity being indicated by modeling, as the first layer of paint was added. As the painting progressed, detail was created in a progressive way, each time by building on the whole. Just as the complex forms of nature appear to be generated through successive additions of smaller and smaller detail, so at one level, a painting could be thought of as growing in a similar fashion.

But of course the generative order of a work of art is far more complex than the preceding description might suggest. For many orders of growth are involved which, in a great painting, are united within a single more comprehensive generative order. The painter may begin with a general idea, a feeling that contains, in a tacit or enfolded way, the whole essence of the final work. The next stage may be to observe the general scene and make sketches that rely upon the sense of visual perception. But in addition to the outward perception, there is also an inner perception in operation which is inseparable from the painter's whole life, training, knowledge, and response to the history of painting. The outward and inward perceptions are, in turn, inseparable from an emotional and intellectual relationship to the theme and even to its literary and social values. Yet this vision is by no means rigid and fixed, for as the painter begins to work on the canvas, a new interaction takes place. He or she is constantly faced with both physical limitations and new potentials, in the very muscular activity of painting and in fresh perceptions of the growing painting beneath the brush.

In all this activity, what is crucial is that in some sense the artist is always working from the generative source of the idea and allowing the work to unfold into ever more definite forms. In this regard his or her thought is similar to that which is proper to science. It proceeds from an origin in free play which then unfolds into ever more crystallized forms. In science as in art it is necessary that what is done with more definite forms should continue at each stage to be open to the kind of free play that is essential to creativity. This holds even if, as with certain artists, such as Matisse, the ultimate form may be a simplification and generalization of what the artist started with, rather than an articulation of greater detail. Matisse's initial creative perception was the constant guide to his activity. This can be seen in the large number of sketches and studies that he made for each of his final paintings, prints, and drawings. His generative idea was clearly the motivation for a subtle and meaningful simplification of lines and forms.[4]

While the essence of the generative order of a painting ultimately escapes definition, it is clear that this order is very different from that of a machine, in which the whole is built out of the parts (i.e., in which the whole emerges through accumulation of detail). By contrast, one of the most important activities during the creation of a work of art is its unfolding, within a particular medium from the original perception. Something similar can be seen in music. Each composition is played

---

4. Simplification in itself is a particularly subtle notion. By reducing detail a painter may, in fact, establish an even more complex order in the work. So-called "simple lines" may create a highly charged dynamic relationship within the painting and its frame. The rhythms of this surface activity will result in a very high perceptive order on the part of the viewer.

in sequential, temporal order, yet its generation can never take place completely within such a sequential way. For that matter the unfolding of the meaning of the music in the mind of a perceptive listener is never totally sequential. This is especially clear in the work of Mozart, who is said to have seen a whole composition in a flash and then to have unfolded it by playing it or rapidly writing it down. Beethoven, by contrast, does not appear to have conceived his works directly as a whole in precisely this fashion, for his notebooks contain themes and sketches worked over long periods of time. Nevertheless, the basic activity in Beethoven's creative work is clearly still a constant unfoldment from a general notion of order.

Bach, for his part, appears to have comprehended fairly directly and as a whole the potential contained within a theme a few bars long, as the following story, told by his son Wilhelm Friedermann Bach, indicates:

> After he had gone on for some time, he asked the King to give him a subject for a Fugue, in order to execute it immediately without any preparation. The King admired the learned manner in which his subject was thus executed extempore; and, probably to see how far such art could be carried, expressed a wish to hear a Fugue in six Obligato Parts. But as it is not every subject that is fit for such full harmony, Bach chose one himself, and immediately executed it to the astonishment of all present in the same magnificent and learned manner as he had done that of the King.[5]

But on returning to Leipzig, Bach was to accept the King's challenge and compose a six-part fugue, nine canons, and a trio sonata on the Royal Theme which he submitted, along with his original fugue, as a Musical Offering. Clearly, in some implicit way the potential of Bach's magnificent composition was perceived by him as enfolded within the King's theme.

There is evidence that in speech the whole meaning is similarly generated quite quickly, along with the language

5. Quoted in H. T. David, *J. S. Bach's Musical Offering*, Dover, New York, 1972.

needed to express it, which comes out as a sequence of words. What is said at any given moment, for example, has never been said in exactly the same way before. In this sense the generative order of language is creative and bears a relationship to artistic and musical creation.

A major feature of a generative order is that through it a process of creation may begin from some broad encompassing overall perception. There is a clue from our language, for the word *generate* has the same root as *general* and *genus*. This supports the earlier claim that, in the arts, creative generation is basically from some general perception, which is then unfolded into particular forms. These may move toward greater and greater detail or, as is the case with Matisse, toward an expression of the general.

FOURIER ANALYSIS

In moving between two extremes, such as art and mathematics, the aim has been to suggest the universal and pervasive character of generative order. For the moment, however, the mathematical side will be stressed, by considering Fourier analysis. For by means of Fourier analysis, a particular arbitrary form can be built out of sets of periodic waves, each of which is of a global order.

Consider such a single wave:

This wave is defined by an order which is similar to itself from period to period. It represents, for example, a wave on a string stretched out in space, or a wave evolving in time. Clearly its order is *global* in that it repeats itself in a similar way indefinitely.

Now add to the first wave a second of double the frequency:

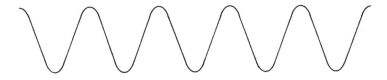

Adding the two together produces:

The diagrams show how more and more waves can be added together to create shapes of any form whatsoever. While each simple wave represents a *global order*, when they are put together they add up to produce a complex *local order* as well.

It is possible to create a well-defined figure in this way based on a generative order which relates the waves of successive frequencies together. This indeed is just how a Fourier series is constructed, for any complex figure can be generated, given a series of coefficients which determine the ways in which the global waves are to be related together. As an example of a Fourier summation, consider a music synthesizer in which a series of oscillators each produce a wave of given frequency, a pure tone. The characteristic sound of any instrument, with all its complex local order, can then be generated by turning appropriate dials on the machine and thereby adding different pure tones together. (In fact a synthesizer also adds characteristics for the attack and decay of each note.)

GOETHE'S URPFLANZE

The simple example of a Fourier series demonstrates how a local order may basically follow from a global order, a reversal of the normal point of view in which global order is regarded as the outcome of local order. But as pointed out earlier, generative orders, especially of a global nature, have so far not been used extensively in science. It is interesting to note, however, that Wolfgang Goethe seriously investigated such a notion two centuries ago. In considering the relationships between the many varieties of plants there are in the world, and the variations that exist within a particular family and genus, Goethe was led to the notion of the *Urpflanze*. Literally this means an original plant and may, at first sight, appear to anticipate Darwin, who envisioned the proliferation of forms as arising through the evolution of original plants and animals. Goethe, however, was not referring to a particular physical manifestation of an original plant but rather to a generative principle or movement from which all plants of a particular kind emerged. There could, therefore, be no actual concrete example of an *Urpflanze*.[6]

Goethe's idea was that this generative principle is subject to a series of transformations of form, a dynamic movement as it were, and that the actual physical manifestations are plants of different but related shapes and features. For example, Goethe considered plants within a particular genus and arranged them as shown in the figure. The various forms of this plant are all related by similar differences. Therefore, the generative principle which, according to Goethe, gives rise to the whole set of plants contains an order of forms implicit within it. Such an ordered set of forms related by similar differences can also be obtained, for example, from a fractal generative process, if a series of small changes in the parameters determining the basic fractal order are made.

6. Ronald H. Brady, *The Causal Dimension of Goethe's Morphology*, in Amrine, Zucker, and Wheeler, eds., *Goethe and the Sciences: A Reconsideration*, Reidel, Dordrecht, Holland, 1986.

These diagrams, taken from Goethe's original researches on morphology, clearly illustrate the nature of the dynamical movement inherent in the urpflanze. *Two particular leaves are linked by a series of transformations that originate from a deeper generative movement.*

Because most of the prevailing ideas concerning the development of form were, at that time, expressed in terms of Euclidian geometry and sequential order, Goethe's notion found little resonance in the science of his day. Nevertheless, perhaps with the development of new ways of looking at such questions, the time has come to explore such directions more fruitfully.

It would seem reasonable, and probably Goethe would have agreed, to suppose that the particular generative order described above is part of a still higher generative order of a wider range of species of plant, going on ultimately to the generative order of plants as a whole and even perhaps to life as a whole. In this way, the generation and evolution of life are thought of as more like the work of an artist than of an engineer. Moreover, considerations of this nature involve a fundamental change in the whole notion of what is meant by a hierarchy of orders. At present the word *hierarchy*, whose root meaning is a government by priests, contains the tacit idea that lower parts of a hierarchy are dominated by higher ones. But in the spirit of generative orders it is possible to consider hierarchies in quite a different sense. Moreover, the inclusiveness of orders, one within the other, is no longer a mere abstract subsumption in the sense that a more general category contains its particulars. Rather the general is now seen to be present *concretely*, as the activity of the generative principle within the generative order. This suggests a new notion of hierarchy, in which the more general principle is immanent, that is, actively pervading and indwelling, not only in the less general, but ultimately in reality as a whole. Emerging in this fashion, hierarchies are no longer fixed and rigid structures, involving domination of lower levels by the higher. Rather, they develop out of an immanent generative principle, from the more general to the less general.

The novel, discussed in the previous chapter, is an example of such a hierarchy, for it grows out of a basic generative order within the author's mind through the generative suborders of plot, character, atmosphere, means of expression, and so on. In addition, this generative order must be expressed within various conventional forms of syntax as they apply within the sentence, paragraph, and chapter, and to the tacit conventions of the novel. Therefore, while within a particular sentence the orders of syntax and semantics may appear to dominate, they are in fact serving the much larger generative order of the novel as a whole. In turn, this larger generative order must serve the aesthetics of style, flow, and readability. So the complex hierarchical order that can be analyzed within a novel is never truly fixed. In a great work of art, it is dynamical and always used in a creative fashion.

Orders in Art

The nature of this dynamical hierarchy of orders, and the generative order that gives rise to it, can also be seen within paintings and musical compositions, and in our relationship to them. For the sake of simplicity, consider a drawing which is composed of lines of varying length, shape, and density made on a piece of paper. At one level the drawing can be apprehended as a series of marks, without any attempt being made to understand or interpret their meaning. The viewer becomes aware of the various orders that are created within this pattern of marks, the symmetry and balance they achieve on the paper as a whole, their rhythms and movement. The marks contain the potential for pattern, repetition, and imitation; that is, for an order of differences and similarities that appear as the eye moves across the field of the paper. In addition, variations and internal differences in their individual structure are also important. Their speed, changing thickness, and means of attack on the paper itself can give rise to complex emotional and aesthetic responses: to feelings of tension, sadness, energy, beauty, and so on. Clearly at the surface level of the marks themselves, with their symmetries and patterns, many orders are involved, with each individual line participating in several different orders and, in the context of each order, bearing differing relationships with its neighbors.

But the surface order of the lines is only one aspect of a drawing, for if it is representational, each line has a particular meaning that can be interpreted as being, for example, part of a body, foliage, a building, or as helping to indicate the solidity of a form or the effects of light. In addition, the relationships between the lines enable the artist to convey a sense of three-dimensional space on a flat surface. Therefore, as well as the complex orders of the surface itself there are a host of additional orders that involve the representation of space, movement, and solid objects. Again, each individual line participates in many different orders, each qualifying and complementing the others in subtle and complex ways. When all these orders are integrated through a comprehensive generative order, a great work of art results, but where integration is only partial or fragmentary then a mediocre work results.

The Baptism of the Ethiopian Eunuch. *Rembrandt van Rijn. Red pen and bistre, whitewash. The National Gallery of Canada, Ottawa. This drawing may be appreciated on many complex levels, through its rhythm of lines, its composition, the humanity of its story, and the skillful way in which it portrays a rich, natural, three-dimensional world. Each stroke of the pen encodes important information about the scene as it describes vegetation, texture, distance, light, and shade. The figures riding across the bridge, for example, immediately establish the necessary scale to place the bridge in the middle distance.*

What is particularly interesting about responding to a drawing or painting is the way in which the viewer interprets, decodes, and responds to this complexity of orders. For in different historical periods and in different cultures, marks and interpretations are made in different ways. The art critic Ernest Gombrich refers to what he calls *schema*, an idea not too dissimilar to Kuhn's paradigms, which take the form of tacitly accepted conventions employed in the construction and viewing of a work of art.[7] Within a given period, most artists employ particular schema, which are then absorbed unconsciously by the viewing public, who interpret the meaning of their works. When a school of art changes, the particular schemata are also transformed and the public may, at first consider the new work to be ugly, meaningless, or "wrong." Until the new schema have been absorbed, the public are unable to interpret, and integrate with their inner perceptions, the work that stands before them. In one sense, therefore, an appropriate syntax has to be employed in reading the painting, just as a syntax is required to read and understand a novel or other piece of writing. In Chapter 3 a similar response was discussed with regard to modern music.

Just as a paradigm is realized when scientists develop fixed habits of mind that leave them insensitive to subtle changes and overemphasize certain obvious differences, so in a similar way can the artist and the viewer become rigid in their responses. Generally it is believed that some "genius" must come along and develop new ways of painting which then enable the public to "see" in exciting new ways.

A drawing or a painting contains many orders that operate together in a dynamical fashion as the work is being made and, for that matter, as it is being viewed. A particularly important aspect of this order is based on the notions of geometry used in its composition. Classical paintings can often be analyzed in terms of simple geometric forms, such as intersecting lines, triangles, rectangles, and circles, that are balanced and arranged in a harmonious fashion. Gestures made with the arms and glances of the eyes, if continued across the painting, will be found to make up one side of a geometric figure which is completed by other gesture lines or a spear, thigh, tree, cloak, or pillar. In addition, the invention of perspective by the architect Brunelleschi gave to painters, beginning with Masaccio, the possibil-

7. Ernest H. Gombrich, *Art and Illusion*, Bollingen Series, Princeton University Press, Princeton, 1972.

ity of a linear order generated by the receding lines and planes of buildings and even of the human body. In a sense this underlying order, which gives structure to many Renaissance paintings, is similar to what we have called the Cartesian order: that is, the underlying use of a grid to portray space and, in the case of a painting, the tacit backdrop on which buildings, people, boats, rivers, and roads are ordered. It is not beyond the bounds of plausibility to see something of a Newtonian order also being anticipated in these Renaissance paintings.

On the other hand, the overwhelming passion of J. W. Turner both in his paintings and poetry was the power of light and the movement of water, so that the underlying order of his art became a form of swirling motion or gyre. In addition, by borrowing from and going far beyond Goethe's theory of advancing and receding colors, Turner was able to give the impression of a constantly rotating vortex within his paintings, a vortex of light, or of the violent motion of air and water that dissolves linear forms.

While Turner's paintings are, of course, important for several reasons, one particularly striking aspect is the way in which the painter was able to overcome the old orders of geometrical structure through the power of his new vortexlike order of light, air, and water in constant motion. It is curious to note that these paintings were made some three decades before J. C. Maxwell published his electromagnetic theory of light, which replaced the Newtonian order of linear trajectories and rigid forms with fields in constant motion and internal rotation. In Turner's "Regulus," reproduced here, it is almost possible to see a new order of movement in which light and air replace the old rigid, linear structure. According to legend, the Roman general Regulus was blinded by the Carthaginians, who cut off his eyelids and forced him to stare at the sun. Turner's painting is created from the perspective of Regulus himself. Around the general is a geometrical order of ships and buildings which are in the process of being dissolved by the blinding sun, whose light radiates from the center of the canvas to cover sea, ships, sky, buildings, and people alike. The paint-

Regulus *(1828, reworked 1837). Joseph Mallord William Turner. The Tate Gallery, London.*

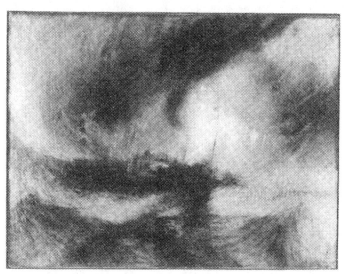

Snow Storm—*STEAMBOAT OFF A HARBOR'S MOUTH MAKING SIGNALS IN SHALLOW WATER AND GOING BY THE LEAD. THE AUTHOR WAS IN THIS STORM ON THE NIGHT THE ARIEL LEFT HARWICH. Joseph Mallord William Turner, The Tate Gallery, London.*

Toward the end of his life Turner painted from a generative order that involved a violent swirling motion, or gyre, of water, light, and air. This black-and-white reproduction does not convey the artist's use of color which also became part of this general movement. **169**

ing seems, therefore, to symbolize a movement toward a new order in art that at least tacitly and implicitly aims to replace the old.

Fresh generative orders, with their hierarchies of line, form, movement, and color require the viewer to respond in new and creative ways that are, for this reason, disturbing. The first attempts the Impressionists made to exhibit their paintings were met with considerable ridicule and critical attack. For in place of the traditional orders and schema of nineteenth-century French painting, Monet had begun to use spots of primary color in an attempt to express his perception of nature through a new way of re-creating, on the canvas, a sense of the order of space. Thus, if you stand close to such a painting, you become aware of the pattern and strength of the color and of its apparent lack of representational form, but as you step back, a whole world with its three-dimensional order seems to come into being. For the viewers of nineteenth-century Paris, however, this use of a new generative order was so different from everything that had gone before in art that it proved to be, initially at least, totally unacceptable to most viewers.

This sort of thing has also happened with scientific revolutions (e.g., relativity). However, just as with scientific revolutions, it turned out that the degree of change had been overestimated, and that the Impressionists had really preserved much of what went before, while making subtle changes in what appeared similar. Corot and Rousseau had also painted in the open air, Constable had loaded his brush with pure white, Delacroix had placed spots of primary color one against the other, and Turner had anticipated Monet's discovery that light dissolves form. So during this whole earlier period, which might perhaps have been called one of "normal art," by analogy with "normal science," a succession of fundamental changes was already taking place. With Monet, however, the change was finally so great that it had to be acknowledged that something really different had appeared, and suddenly people began to feel that they were in the midst of a "revolution in art."

The process was carried further by later artists. Thus to

Cézanne, Monet was "just an eye, but my God what an eye."[8]
Cézanne went on to transform Impressionism by radically exploring composition and the structure of objects and land-
scapes while still preserving some of the Impressionistic order.[9]
Cézanne's contribution is particularly apparent in his paintings
of Mount St. Victoire, which are highly organized globally
according to various planes. Indeed, Cézanne's generative order is reminiscent of the Fourier order described earlier. In its
most extreme form, this new order was used by Picasso and
Braque in their Cubist paintings but can also be found, in
more subtle ways, in many other schools of twentieth-century
art.

What then lies in the creation of a new form of art and in the
viewer's ability to perceive it?

Each artist creates using a generative order which contains a
highly complex and dynamic hierarchy of orders of line, form,
color, meaning, and so on. While the mediocre are content to
pursue their habits of mind and do not have sufficient passion
and energy to create beyond the generative schema that went
before, the great artist is able to perceive the world in new ways
and to create fresh orders within his or her paintings. Likewise
the viewer who is both passionate and sensitive will be able to
explore new generative orders within the mind and respond to
the various clues that are present on paper or canvas. Looking
at a work of art is a creative act which leads to an order similar
to that which the artist had in mind when the original work was
created. In this way it can be truly said that an artist teaches
us to see the world in new ways. For the activity of reading and
understanding a work of art involves the creative perception of
new generative orders which ultimately lie beyond that individual work and extend to the whole of nature and experience.

8. Ambroise Vollard, *Cézanne*, Dover, New York, 1984.

9. An instructive account of this transformation, along with its background,
is contained in C. Biederman, *The New Cézanne*, Art History Publications,
Red Wing, Minn., 1958. A structural analysis of the paintings is made in
Erle Loran, *Cézanne's Compositions*, University of California Press, Berkeley,
1985.

THE IMPLICATE OR ENFOLDED ORDER

In science as in art, it is necessary to allow for the emergence, in creative perception, of new generative orders, which go beyond the individual content and involve the whole, common cultural experience. It is our suggestion that the implicate or enfolded order has such a potentiality. This form of order, which has been treated by David Bohm in *Wholeness and the Implicate Order,*[10] is in a close relationship to that of fractals in the sense that, in both, there is a kind of whole generated from certain basic principles. Nevertheless, the implicate order extends the notion of generative order beyond what can be done with fractals. For this reason a detailed discussion of this form of order is given in this section. It will provide a powerful tool for exploring the ideas of creativity and generative order later on in the book.

The implicate order can be illustrated with the aid of the following device: Consider two concentric glass cylinders, the inner one fixed and the outer capable of being slowly rotated. The space between the cylinders is filled with a viscous liquid such as glycerin. When the outer cylinder is turned, fluid close to it is dragged along at nearly the same speed, but fluid close to the inner, stationary, cylinder is held nearly at rest. Hence fluid in different regions of the space moves at different rates, and in this way, any small element of the glycerin is eventually drawn out into a long thin thread. If a drop of indissoluble ink is placed in the liquid, then it becomes possible to follow the movement of the small element by watching how the drop is drawn out into a thread until eventually it becomes so fine as to be invisible.

At first sight one may be tempted to say that the ink drop has been totally mixed into the glycerin so that its initial order has been lost and is now random or chaotic. But suppose that the outer cylinder is now rotated in the reverse direction. If the fluid is very viscous, like glycerin, and the cylinder is not rotated too quickly, then the fluid element will in fact retrace its steps exactly. Eventually the element will return to

10. Routledge and Kegan Paul, London, 1980.

its original form and the droplet of ink will appear as if from nothing. (Such devices have in fact been constructed and the effect is indeed quite dramatic.) Clearly what was taken for a chaotic or random loss of order was in fact a hidden order of high degree that was generated out of the initial simple order of the drop by means of the turning cylinder. Likewise this hidden order was transformed back into the original simple order when the cylinder was reversed. Clearly, there is a correspondence between this hidden order and the discussion in the previous chapter of how hidden order may quite generally be contained in what seems to be chance or randomness.

While the drop is present in hidden form, it may be said to be enfolded into the glycerin rather as an egg may be said to be folded into a cake. When the cylinders rotate in the reverse direction, the droplet then unfolds. With the egg in the cake, this is of course not possible, the reason being that the fluids in the cake are not sufficiently viscous.

To obtain an enfolded or implicate order from such hidden orders it is, however, necessary to consider a whole series of ink drops, enfolded in succession. Let us put in the first drop and rotate the cylinder n times. A second drop is now added and likewise enfolded n times, which also has the net effect of enfolding the first drop $2n$ times. A third drop is added and enfolded n times, the second being enfolded $2n$ times and the first $3n$ times. This process is repeated until many droplets have been enfolded. When the cylinder is reversed, one droplet after another will emerge into an unfolded or explicate form and then fold back into the glycerin again. If this is done rapidly, the overall effect will be of a droplet which appears to subsist for a time within the moving liquid.

The experiment can now be extended so that the droplets are added in successively different positions. When the cylinder is reversed, the droplets will unfold in a line that moves across space. If the movement is rapid enough, this will give the impression of a particle that crosses space along a trajectory. However, this particle is simply a manifestation of a much greater enfolded or implicate order within the whole of the glycerin, most of which is hidden. As has been explained in

Wholeness and the Implicate Order, this provides a good analogy to many of the basic quantum properties of particles, such as electrons. For example, the ink droplets may be so arranged that they produce a continuous track up to a certain point which then jumps continuously to start another track at a finite distance away, thus providing a way of understanding discontinuous "jumps" of the electron from one quantum state to another.

The above example helps to indicate what is meant by the implicate or enfolded order. What is essential to such an order is the simultaneous presence of a sequence of many degrees of enfoldment with similar differences between them, for example, the ink droplets in the glycerin. Such an order cannot be made explicit as a whole, but can be manifested only in the emergence of successive degrees of enfoldment. This may be contrasted with an explicate or unfolded order, in which the similar differences are all present together, in a manifest and extended form. This explicate order is of course commonly found in ordinary experience and in classical physics.

It is clear that the explicate order corresponds very well to a worldview in which the basic notion is one of separate objects moving on trajectories. These trajectories, in turn, can be described in terms of Cartesian coordinates, as was indicated in Chapter 3. Although physics has gone through a revolution in which the notions of particle and trajectory have ceased to be basic, the underlying Cartesian coordinates still pervade the mathematical formalism. And thus, the mathematics has hidden within it a key survival of the old order. This may well account for some of the difficulties that science has in connection with relativistic theories, both special and general. The implicate order, however, has the possibility of opening up very different approaches in which these difficulties may perhaps not arise.

Analogies like that of the ink drop are, however, limited because the actual particles that make up the ink droplet themselves move in an explicate way, even though the development of the droplet itself may be quite complex. A better analogy to the behavior of an electron, for example, can be obtained by considering a holograph, which is a photographic record of light waves that have been reflected from an object.

In normal photography a lens is used to focus light from an object, so that each small section of the object is reproduced in a small section of the photographic plate. In holography, however, the photographic record made by laser light does not in fact resemble the object but consists of a fine pattern of interference fringes. Each portion of the plate now contains information from the whole of the object. When similar laser light is used to illuminate the plate, the light waves emerging from it resemble those that originally came from the object. It is therefore possible to see, in three dimensions, an image of the original object. What is particularly significant, however, is that even if only part of the plate is illuminated, an image of the whole object is still obtained. This is because light from every part of the object is enfolded within each region of the plate. In normal photography, information is stored locally, but with the holograph it is stored globally. As successively smaller regions of the holograph are illuminated, the images as a whole are not lost. Instead fine detail becomes progressively more difficult to resolve. This global property of enfoldment of information and detail has something in common with both fractal and Fourier orders.

The holograph provides a good analogy to the general nature of movement according to quantum mechanics. This movement is described mathematically by what is called a Green's function, which can be thought of as representing a summation of very many waves, similar in some sense to those that scatter off an object. It is possible to obtain an intuitive picture of the meaning of the Green's function by considering what is known as a Feynmann diagram. This is a representation of the movement of waves in terms of a diagrammatic structure of lines.

To start, consider a wavelet that emanates from a fixed point P:

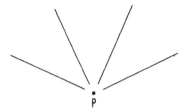

The lines radiating from P show how the wavelet spreads from this point. Now consider any point Q, at which the wavelet from P arrives. This in turn is the source of another wavelet, which spreads as follows:

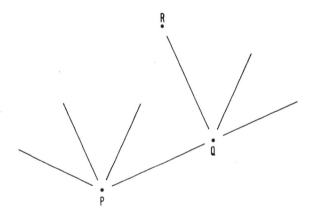

In this way the point R is reached, which itself becomes the source of yet another wavelet and so on. The essential idea is that each point is reached by wavelets, from all other points. In turn it becomes the source of a wavelet, proportional in strength to that of the wave which reaches it. Evidently there is a basic notion of order here, because the differences in any given step are similar to those in the next, and so on.

    Now look at this process as a whole, and begin by considering all those waves, emanating from A and arriving at B, after a large number n of intermediate steps.

Below is a typical path that connects A and B.

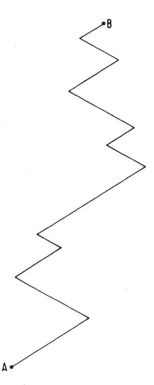

The total wave at B is the sum of the contributions of all possible paths of this kind that connect A and B.

This way of looking at wave movement was proposed originally by C. Huygens in the eighteenth century, but around 1950, R. P. Feynmann developed his diagrammatic representation of the approach of Huygens. The result was a very powerful new tool for dealing with quantum mechanical field theory.

At first Feynmann did not intend this simply to be a tool, for he hoped that it would provide physical insight into quantum processes. Indeed, the informal language used in connection with Feynmann diagrams, i.e., that they are the definite "paths" of particles, would suggest that such physical insight is possible. However, these diagrams actually represent only the contributions of different wavelets, which may add or subtract to produce interference effects, and so they are not coherent with the idea that a particle actually follows such a path. Since the electron not only is a wave but also has a particle nature, the Feynmann diagrams cannot provide an adequate image of the actual movement from whichever standpoint they are regarded.

What was left to physicists therefore was only to look at these diagrams as being an extremely useful tool in performing certain difficult calculations. However, it may turn out that Feynmann's original intuition may have some further meaning which has not yet been seen. Perhaps the implicate order will be relevant to seeing this.

Returning to a consideration of the implicate order: Evidently the Feynmann diagrams give an imaginative picture of a wave motion. In this picture, wavelets can be seen unfolding from each point toward the whole. Yet the very same movement can also be thought of as wavelets enfolding toward each point from the whole, as shown in the diagram.

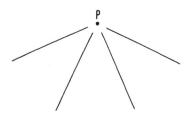

The basic movement of enfoldment and unfoldment is thus a dual one in which there is ultimately no separation between enfoldment and the unfoldment. The movement has the order of similar differences of degrees of enfoldment and unfoldment that has already been indicated. It therefore provides an example of the implicate or enfolded order, which is described mathematically by the Green's function and graphically by the corresponding Feynmann diagrams.

Clearly this interpretation of the Green's function is ultimately an outcome of physical intuition, on the part of Feynmann, so that the source of the ideas is not primarily in the mathematics. In quantum theory all movement is described in terms of Green's functions in the way indicated above. It follows therefore that the basic movements and transformations of all matter and all fields are to be understood in terms of a process of this kind. It is even possible to obtain some fairly direct experience of how it takes place by considering that as a person sits in a room, light from all points in it must enfold together to enter the pupil of the eye. This complex information is then unfolded by the lens of the eye and the nervous system into a consciousness of the room as constituted out of extended objects in an explicate order. Indeed there is even evidence that the memory of this event will not be stored locally within the brain but is distributed in some global fashion that resembles the implicate order.

More generally, with a telescope, the entire universe in space and time is enfolded within each region and can then be unfolded with the aid of lenses and cameras. At first sight it could be supposed that the light coming from all the stars would produce a totally disordered pattern of waves within any small region of space. Yet each region enfolds the whole universe. Indeed, it is just this process of enfoldment and unfoldment that allows scientists to learn about the whole of the universe, no matter where they may be in it.

In the usual way of thinking, something like an implicate order is tacitly acknowledged, but it is not regarded as having any fundamental significance. For example, processes of enfoldment, such as those described by the Green's function, are assumed to be just convenient ways of analyzing what is basically a movement in the explicate order, in which waves are transmitted continuously through a purely local contact of fields that are only infinitesimal distances from each other. In essence, however, the main point of the implicate order is to turn this approach upside down, and to regard the implicate order as fundamental, while the explicate order is then understood as having unfolded from the implicate order.

This has been illustrated through the analogies of the ink droplets and of the hologram. It is possible to combine certain features of both these analogies by imagining a wave that comes to a focus in a small region of space and then disperses. This is followed by another similar wave that focuses in a slightly different position, then by another and another and so on indefinitely until a "track" is formed that resembles the path of a particle. Indeed the particles of physics are more like these dynamic structures, which are always grounded in the whole from which they unfold and into which they enfold, than like little billiard balls that are grounded only in their own localized forms.

It is necessary, however, to go further than this. Up to now particular kinds of entities, such as electrons and neutrons, have been discussed, each of which has its own implicate order. But there may be a further unknown set of entities, each having its implicate order, and beyond this there may be a common implicate order, which goes deeper and deeper without limit and is ultimately unknown. This unknown and undescribable totality will be called the *holomovement*. It acts as the fundamental ground of all matter. As in the case of the analogy, in which a particle is taken to be a succession of wave pulses, so each object or entity emerges as a relatively stable and constant form out of the holomovement and into the explicate order. This form is sustained by the holomovement, into which it eventually dissolves. Therefore it must be understood primarily through this holomovement. It is clear that the implicate order ultimately prevails, although it is always in an essential relationship with the explicate order.

THE SUPERIMPLICATE ORDER

The discussion based on the hologram provides only a limited view of the implicate order because it is based on a classical treatment of the transformations within a light wave. To obtain a deeper and more extensive understanding of the implicate order, it is necessary to start from quantum mechanical field theory. This is, in essence, the most basic and general form of the modern quantum theory that is available to date. Such a step will lead to an extension of the implicate order, called the superimplicate order. This is much subtler than the implicate order and goes deeper. In addition, it is capable of further extensions in ways that go beyond quantum theory altogether.

As with the quantum mechanical particle theory, it is necessary to proceed from the standpoint that the mathematical formalism of the quantum field theory is essentially correct, at least within some suitable limits. However, the informal language describing physical concepts is even more unclear in the field theory than it is in the particle theory.

The causal interpretation will therefore be extended in order to obtain a clear physical notion of the quantum field theory, as well as to gain insight into the superimplicate order.[11]

To be more specific, the key new property treated by quantum field theory is the appearance of discrete particlelike quanta, in what was initially assumed to be a continuous field. In certain ways this process is similar to what was described in the analogy of a wave that focuses in a succession of small regions and thus resembles the track of a particle. In other ways, however, it is quite different so that this analogy, too, is limited. But in the causal interpretation, a clear and well-defined physical concept of the appearance of discrete particlelike quanta in a continuous field can be given. This interpretation fully and faithfully expresses the meaning of the mathematical equations.

It must be emphasized, however, that although the particular example of the superimplicate order is obtained from the causal interpretation of quantum field theory, the essential idea of the superimplicate order is not restricted either to the causal interpretation or to the quantum theory itself. Rather, these are only special forms of the more general superimplicate order.

The basic discussion of quantum field theory in terms of the implicate order and the causal interpretation is quite simple.

11. For a more detailed treatment, see D. Bohm and B. Hiley, *The Causal Interpretation of Quantum Mechanical Field Theory*, forthcoming.

Instead of taking a particle as the fundamental reality, start with the field. And instead of having a particle acted on by a quantum potential, suppose that the field is acted on by a superquantum potential. This superquantum potential is far subtler and more complex than the quantum potential, yet the basic principles governing its behavior are similar. Its net effect is to modify the field equations in a fundamental way so they become nonlinear and nonlocal. This brings about the new quantum properties of the field.

The field is continuous and by itself would tend to spread out from any source. However, because the superquantum potential is nonlinear and nonlocal, it is able, under certain conditions, to provide a very subtle kind of immediate connection between distant regions of the field. Suppose, for example, that the field meets an atom that can absorb a definite amount of energy. The superquantum potential will "sweep in" energy from the whole field, in a definite amount equal to what can be absorbed by the atom. This explains how a continuous field can act in matter as if it were made up of discrete elementary units.

In the particle treatment, the wave-particle duality was explained as an effect of the quantum field on the particle. But the wave-particle duality can now be treated as an effect of the superquantum "field" on the original field itself. Therefore, the particle is no longer used as a basic concept, even though the field manifests itself in discrete units, as if it were composed of particles.

So far the implicate order has not been brought in. Indeed, in the particle theory, the causal interpretation, with the prominence given to the quantum potential, appears, at least at first sight, to be a step away from regarding the implicate order as basic. But in the causal interpretation of the field theory, this is not so. Indeed, in this case there are *two* implicate orders in a specified relationship. The *first implicate order* is just the field itself, and its movement, as described by Green's functions, is just a form of the implicate order. The *second implicate order* is then obtained by considering the superquantum wave function. This is related to the whole field as the original quantum wave function is related to the particle. A more detailed treatment shows that the superquantum wave function also moves in a kind of implicate order which is, however, far subtler and more complex than is the first implicate order. This then comprises the second implicate order.

In the earlier version of the causal interpretation, given in Chapter 3, the quantum potential represents information which guides the self-active movement of the particles. In the field case, the superquantum potential now represents information that "guides" or organizes the self-active movement of the field.

The first implicate order applies to the original field which, however, now has new features brought about by the action of the superquantum potential. And the second, or superimplicate, order applies to the "superfield" or information that guides and organizes this original field.

A good analogy to the first and second implicate orders is provided by considering a computer or video game. The first implicate order corresponds to the television screen, which is capable of showing an indefinite variety of explicate forms, which are essentially manifestations of an implicate order. In earlier television sets this could clearly be seen through the action of the synchronizing adjustment. When synchronism failed, the images would be seen to enfold into an apparently featureless background. But when the correct adjustment was made, the hidden images would suddenly unfold into explicate forms again.

The second implicate order corresponds to the computer, which supplies the information that arranges the various forms—spaceships, cars, and so forth—in the first implicate order. Finally the player of this game acts as a third implicate order, affecting the second implicate order. The result of all this is to produce a closed loop, from the screen to the player to the computer and back to the screen.

Such a loop is, in a certain sense, self-sustaining, for with only the computer and the screen in operation, all that would happen would be an unfoldment of a predetermined program. But when the player, as third implicate order, is introduced, a closed loop results and the possibility is opened up of a genuine dynamic development in time, in which creative novelty may enter.

We speculate that, in nature, there is something like a third implicate order that affects the second and is affected by the first, thus giving rise again to a closed loop. Or more generally there is an indefinite series, and perhaps hierarchies, of implicate orders, some of which form relatively closed loops and some of which do not. Of course such an idea implies that the current quantum theory is of limited validity. This theory is covered only by the first and second implicate orders. Where anything beyond the second implicate order is active, then quantum theory would no longer be valid.

THE RELATIONSHIP BETWEEN THE IMPLICATE ORDER AND THE GENERATIVE ORDER

The implicate order and the generative order are very closely related. Indeed, the implicate order may be understood as a particular case of the generative order. Thus, in the discussion of the Green's function, it is possible to see how explicate forms are generated in an order of unfoldment, step by step. In this process, the whole is relevant to each part, unlike the case of fractals where the details are generated from local forms belonging only to the next less detailed stage of generation. While the implicate order is similar in a certain way to the order of fractals it is much more extensive and subtle, since the process of unfoldment is related to the whole and not to a local order of space.

There is, however, a much more fundamental sense in which the implicate order is a generative order. For in quantum field theory, and the computer game analogy, the second implicate order is basically the source from which the forms of the first implicate order are generated. If there are higher implicate orders, then a similar generative order will prevail throughout all the levels. Ultimately, it is, of course, the holomovement, and what may lie beyond, from which all is generated.

THE IMPLICATE ORDER AND CONSCIOUSNESS

So far only material processes have been discussed in terms of the implicate order. But consciousness is much more of the implicate order than is matter. This is brought out in some detail in David Bohm's *Wholeness and the Implicate Order*. A few of the main points discussed in that book, along with some further notions on the relationship between mind and matter, will be presented here.

First of all, it is clear that thought is definitely in the implicate order. The very word *implicate*, meaning enfolded, suggests that one thought enfolds another and that a train of thought is actually a process of enfoldment of a succession of implications. This is not entirely dissimilar to the process described by a Green's function, or to what takes place in the video game. In addition, thoughts and feelings unfold into each other, and these in turn give rise to dispositions that unfold into physical actions and on to more thoughts and feelings.

Language is also an enfolded order. Meaning is enfolded in the structure of the language, and meaning unfolds into thought, feeling, and all the activities that have already been discussed. In communication, meaning unfolds into the whole community and unfolds from the community into each person. Thus, there is an internal relationship of human beings to each other, and to society as a whole. The explicate form of all this is the structure of society, and the implicate form is the *content* of the culture, which extends into the consciousness of each person. What is seen on one side as society and the explicate *forms* of culture enfolds inseparably within what is seen on the other side as the consciousness of each individual in the society. For example, the laws and customs and limitations of the society do not actually operate as external forces that are alien to the people on whom they act. Rather, they are the expression of the very nature of these people, and in turn, they enfold to contribute to this nature.

Evidently, the implicate order of consciousness operates on many levels, which are related to each other as are the implicate and superimplicate order of the quantum field, and of the computer game. For example, as in the discussion of reason in the previous chapters, it was shown how one level of thought will organize the next level. This can go on to produce a structure that may develop indefinitely with relatively closed loops of many kinds. This implies that consciousness is organized through a generative order whose totality is in many ways similar to the totality of the generative and implicate order that organizes matter.

It is now possible to look into the question of how consciousness and matter are related. One possibility is to regard them as two generative and implicate orders, like separate but parallel streams that interrelate in some way. Another possibility is that basically there is only one order, whose ground includes the holomovement and may go beyond. This order will unfold into the two orders of matter and mind, which, depending on the context, will have some kind of relative independence of function. Yet at a deeper level they are actually inseparable and interwoven, just as in the computer game the player and the screen are united by participation in common loops. In this view, mind and matter are two aspects of one whole and no more separable than are form and content.[12]

THE EXPLICATE AND SEQUENTIAL LIMITS OF THE IMPLICATE AND GENERATIVE ORDERS

If the implicate and generative orders are fundamental, both to mind and matter, how is it possible to account for the fact that, in ordinary experience, explicate orders of succession appear to dominate? In the causal interpretation of quantum theory, it happens that a simple answer can be given to this question with regard to matter. In Chapter 2 it was explained that the quantum potential becomes negligible in the domain of large-scale experience. To put it another way, what we take to be our own domain of experience is just that in which the effects of the quantum potential can be neglected. A similar

12. The possible connections between mind and matter are also discussed in F. David Peat, *Synchronicity: The Bridge Between Matter and Mind*, Bantam, New York, 1987.

situation also holds for the superimplicate order so that all the subtle effects of the implicate and generative orders do not normally manifest themselves at the level of ordinary (classical mechanical) experience. The behavior of matter, in this limit, reduces either to that of Newtonian particles or classical continuous fields that do not manifest themselves in a "quantized" or particlelike way.

Something similar also happens with regard to thoughts and feelings in the field of consciousness. To see how this comes about, note that explicate orders are in fact simple patterns or invariants in time, that is, aspects which persist or repeat themselves in similar ways and have well-defined locations in space. In other words, these orders are associated with relatively simple orders of similarities and differences. Sense perception, while operating at its deepest levels within a generative, implicate order, tends to abstract that which is relatively invariant or slowly moving against a subtler and more dynamic background. It then deals with the environment in terms of such relatively simple similarities and differences.

Recall the example of human vision, in which the center of the eye selects structures of similar differences against a background of subtler and more dynamic similarities and differences sensed by the periphery of vision. In addition, in walking around a form, its appearance changes in radical ways, both through a change of orientation and as a result of variations in illumination. Perception and consciousness, however, abstracts from this continuous change what is invariant or slowly varying and identifies this as a single solid object. (A discussion of such abstraction, which also compares this process with what is done in physics, can be found in David Bohm's *The Special Theory of Relativity*.[12])

Abstraction of invariants from a deeper implicate order is even more strikingly demonstrated by considering that an individual human face can be identified in a crowd of moving people. A face changes considerably over a lifetime and under varied conditions of lighting, makeup, or facial hair, yet it is

12. Benjamin, New York, 1985.

still possible to recognize the face of a friend who may not have been seen for many years.

In weak illumination, however, there is no clear perception of form. Instead we begin to become aware of a constantly changing sensation of light and shadow, and of how the mind acts within its generative orders in an attempt to incorporate these constantly changing impressions until they fit in a relatively invariant way. Visual perception under these conditions is closer to the original implicate order, as the mind tries to construct something explicate out of the shifting information. The explicate order that it endeavors to build will not be firm at first. With the advent of new information, the experience of a form may suddenly change in a radical way. Only after this process has continued for some time does the explicate form remain stable.

With regard to thoughts, feelings, and other internal mental processes, it is clear that these arise also from an ever-changing and fluctuating background of the stream of consciousness. Most of these are transient, and have little firm definition. Thus it is only through organized thought, which generally takes place in a social and cultural context, that ideas are able to take definite form and to "stand firm." Also, emotions tend to change rapidly, and it is primarily through naming them and forming concepts about them that it is possible to hold them fixed. Moreover, naming an object and forming concepts about it plays a crucial role in giving a definite shape and form to sense perception. This was clearly brought out in the case of Helen Keller.

Finally the implicate order can be experienced directly, not only in connection with the fluctuating background of consciousness but also in relationship to perception of certain kinds of well-defined forms. Consider, for example, how music is comprehended. At any given moment, a particular note may be sounding in awareness, but at the same time, a kind of "reverberation" of a number of earlier notes can also be sensed. Such reverberation is not the same as recollection or memory. Rather it is more like a part of an unbroken enfoldment and unfoldment of the notes concerned into ever subtler forms, including emotions and impulses to physical movement, as well as a kind of "ethereal" echo of the original notes within the mind. Indeed if successive notes are played several seconds apart, then they no longer combine together in such a way as to convey the dynamic sense of unbroken flow that is essential to the meaning of the music. But when they are played at their proper speed, the notes fold together into an overall tune or musical theme.

This suggests that, at any given moment, a number of notes are present in awareness in various degrees of enfoldment. The simultaneous awareness of all of these is what constitutes the sense of unbroken flow that has been described above. But this means that it is possible to be directly aware of an implicate order as a set of similar differences that are present simultaneously in different degrees of enfoldment of successive notes. This corresponds roughly to the simultaneous presence of a set of droplets in similarly different degrees of enfoldment in the glycerin.

On a much greater scale, the perceptions of Mozart and Bach of whole musical structures in single flashes of insight probably involved an order that was not only implicate, in the sense of containing an overall hierarchy of structure all at once, but also generative, in the sense that it contained the overall order out of which it enfolded.

This perception of the implicate order is generally common to all works of art. For example, the montage, or editing together of successive images, in the film of a great director has something in common with music, for the internal structure, quality, and feeling of each image infuses all the others. In this way the value and meaning of a particular image, seen alone, is totally transformed and the resulting scene is viewed as an organic whole rather than as a succession of explicit images. In poetry the various resonances of words and images act together in highly complex orders so that associations of memory and meaning in an individual word or image, together with the particular sounds its vocalization (aloud or in the mind) evokes, are all enfolded together. The act of reading a poem is that of realizing the order of these enfolded forms and attempting to reach the generative order at the heart of the work.

It follows that implicate and generative orders of such kinds are ultimately at the ground of all experience. However, this is generally ignored, probably because societies have generally deemed it to be absolutely necessary for their survival to emphasize explicate orders, which are especially suitable for large-scale organization and technology.

It is clear then that the explicate order of succession, which appears to stand on its own, actually arises out of an organization that lies in the implicate and generative orders and that is never free from the possibility of collapsing as further data appear. The implicate and generative world is clearly the ground of all experiencing, and the explicate world of succession is constructed out of this ground. Through habits of thought and language, people have come to take the explicate world of succession as the true ground and the implicate and generative orders as something that is secondary to such a ground in the explicate world. It is therefore particularly important that this view be turned around in order to understand reality more deeply.

SUMMARY AND CONCLUSIONS

This chapter moved beyond the idea of a sequential order, to introduce the notion of a generative order. The first example was in terms of fractals. Following this, it was shown how the generative order is relevant to creativity in art and to the creative perception and understanding of nature. The next step was to go into the implicate order, showing how it leads to the superimplicate order, which in turn organizes the implicate order. This opened the way for an indefinite extension into even higher implicate orders, which organize the lower ones, while capable of being affected by them. In this way it became apparent that the implicate order is a very rich and subtle generative order. Finally, consciousness was discussed as a generative and implicate order, and through this, notions of how mind and matter are related were proposed.

In the final chapters, this approach will be extended to throw light on nature, mind, and society in a general way. This will help to open the door to a kind of dialogue that may creatively meet the breakdown of order that humanity is experiencing in its relationships in all these fields.

FIVE

GENERATIVE ORDER IN SCIENCE, SOCIETY, AND CONSCIOUSNESS

In the previous chapter the new notion of generative order was introduced. Its close relationship to the implicate order and, by extension, the superimplicate order was shown. In this chapter the generative order will be explored further and its importance in physics, cosmology, biology, and consciousness will be discussed. However, it is not the main purpose of this book to use the generative order to develop new scientific theories, although such possibilities should of course be explored in other contexts. Rather, the study of the generative order is aimed at helping to understand the meaning of creativity and to discover what is blocking it.

As the operation of the generative order is unfolded, in this chapter, it will become apparent that creativity, in its essence, cannot really be divided into different fields of specialization, for it is one whole. Therefore a fundamental and far-reaching surge in science would have to be combined with a similar surge in these other fields and in all areas of life. The discussion of the generative order will, therefore, help to prepare the ground for the final chapter, which explores the question of how such general creativity may be fostered.

GENERATIVE ORDER IN
PHYSICS AND IN COSMOLOGY

A number of reasons have already been given for the proposal that a generative order, in the form of the superimplicate order, lies at the foundation of physics. This was shown explicitly in the case of light. In the first implicate order this is basically a movement of a field, and yet, through the information in the second implicate order, this movement is organized into a particlelike behavior. More generally, all of the so-called elementary particles can now be treated in this way, as quantum mechanical fields that are organized by information in their superimplicate orders which makes possible the creation, sustenance, and annihilation of particlelike manifestations. They are thus relatively constant and autonomous particlelike features of the holomovement that emerge through the generative order.

Such "particles" can therefore, in a wide range of circumstances, be abstracted as relatively stable units. When such simplification is appropriate, the theory reduces to the original causal interpretation in which the wave function, now dependent only on the "particle" coordinates, provides a pool of information that fundamentally affects the behavior of each effective "particle" through the quantum potential. These effective "particles" can be organized in this way into higher-level entities, such as atoms and molecules. In this way the activity of structuring can be seen to be sustained from a deeper generative order. This order organizes the particles in ways that depend crucially on the common pool of information in the wave function of the whole system. This, however, cannot be expressed solely in terms of the particles and preassigned relationships between them. As a result, new kinds of properties can arise, not anticipated in classical theories, in which the feature of quantum mechanical wholeness, which was discussed in Chapters 2 and 3, plays a key part. Such properties pervade the whole of physics and of chemistry as well.

Even to abstract an atom or molecule out of its general environment, as an autonomous entity, is still an approximation. For, in principle, the relevant common pool of information may encompass larger systems and ultimately the whole of the universe. As an example, consider the property of superconductivity in metals, in which quantum mechanical features appear at low temperatures and over macroscopic orders of distance. At higher temperatures the wave function of the whole system breaks up into many independent pools of information, and so the electrons move relatively independently. Each electron is then easily deflected by irregularities in the arrangement of atoms in the lattice, including especially those produced by thermal vibrations in the lattice itself. And in this way the current-carrying electrons are scattered in random directions, with the result that free flow is impeded and electrical resistance results. On the other hand, at low temperatures, the wave function corresponds to a single common pool of information which "guides" electrons and keeps them moving together in spite of the irregularities that would otherwise deflect them and break up the coordination of their movements. The net result is that the current flows without resistance. This situation may be compared to that of a ballet in which all the dancers move and go around obstacles together, according to a common score, which constitutes a single pool of information. At higher temperatures, however, the electrons do not behave in this way but, rather, like a disorganized crowd of people, in which each person is guided independently by his or her own information. As a result, the movements tend to disperse in random directions.

In chemistry, the properties of a molecule also depend critically on the common pool of information which is enfolded in its wave function and which may extend into wider and wider contexts. Therefore, the explicate order cannot give a complete account of the emergence of new chemical properties, just as it cannot account for the emergence of superconductivity. This implies that the general goal of absolute reductionism is not feasible; if by this is meant the proposal that *all* properties of matter can be explained through explicate structures alone. This limitation arises because the generative and implicate orders are involved in an essential way, even at this rather elementary level of physics and chemistry.

In Chapters 2 and 4 it was shown that there is a limit in which the contributions from the quantum potential can be neglected. In this limit ordinary classical (Newtonian) motions provide a good description; on the whole they give a good approximation to the domain of common, large-scale experience. Nevertheless under special conditions, for example, extremely low temperatures, the quantum potential may be significant even on the large scale. Superconductivity and superfluidity, for example, are large-scale manifestations of the effects of the quantum potential. More generally, this potential plays a significant role in many other ways. Without it, it would not be possible to account for the stability of atoms and the chemical properties of molecules that constitute bulk matter. Nor could many of the basic properties of solids, such as crystals and metals, be explained. In addition, the possibility must be kept open that in macromolecules, and in other contexts, certain properties, as yet unexplained, may be very sensitive to the quantum potential. But, it must be emphasized, in spite of all this a suitable classical limit does exist in which explicate forms appear to dominate. This explains why generative and implicate orders are not significant in the broad areas of experience in which classical physics is valid.

The causal interpretation has been discussed as a particular way of describing the generative order in physics. But even if this particular theory is sooner or later replaced, the essential idea of the generative order, which operates through various levels of information, will still be relevant. Indeed, throughout this book it is being suggested that such a generative order goes far beyond the quantum theory and is a key feature of the general notion of order that is relevant for understanding creativity in all areas of life.

When it comes to a consideration of the cosmos, it is clear that the current big bang theory, along with its various developments, has produced a great many interesting results. However, it also contains a number of serious difficulties. Among these, one of the most problematic, which was referred to in Chapter 2, is the collapse of the wave function. This problem is particularly serious near the moment of origin of the universe, where there can be neither observers nor instruments which make measurements, and so bring about the "collapse of the wave function." Nevertheless the usually accepted interpretations, including the many worlds approach of Everett, all require either the assumption of measuring instruments or, as in Wigner's interpretation, an observer outside the material universe in the form of a pure disembodied spirit, in order to give physical meaning to the mathematical equations.

In the present approach, however, there is no need to bring in such assumptions, which in any case are extraneous to the mathematical laws of quantum theory. Rather, the universe can be discussed as grounded in *that which is*, out of which emerges an overall reality that also includes subjects, who can act as observers.

As was suggested in Chapter 3, each subject can be considered as a microcosm, who stands in relation to the whole as an inexhaustible source of analogies. The observer and the observed are thus internally related by a totality of "ratios" or proportions, which are enfolded in both. This relationship can then unfold in the cycle of perception-action between the two. There is therefore no need, as in the usually accepted interpretations of the quantum theory, to depend on the assumption of an observer to give meaning to the theory. Rather, the whole process of the universe can be discussed in an approach that does not require a fundamental distinction between observer and observed. For this reason, it is not necessary to become involved in the various complicated and arbitrary assumptions that have been proposed about the nature of observers in which the other interpretations are entangled.

In a discussion of the "wave function of the universe," the implicate and generative order of the whole is involved. Out of this whole emerge subwholes (including observers), which themselves can be organized into wider contexts, and these in turn are further organized until the total universe is encompassed in this way. As far as the interpretation of the present status of physics is concerned, this generative order need be carried only from the first implicate order to the second (i.e., the superimplicate order). But in accordance with the proposals of this chapter, this is itself an abstraction from a much vaster generative order, with a complex structure of relatively closed loops, which can in principle go on indefinitely to ever greater depths of subtlety. Current physics has only "scratched the surface" of this order so far. For example, in a deeper study that would go beyond the current quantum theory, intimate connections in the implicate and generative orders may be discovered between different scales of time, mass, energy, and distance, and a complex web of relationships between all these different scales may be found. Already such correlations between different regions of space have been found through the Einstein, Podolsky, and Rosen experiment. It is now possible to speculate that such scales may stand in the relationship of analogies to each other, with similar forms and proportions in different contexts. In general, such relationships will be of an essential nonlocal nature. In this way, the universe could exhibit new kinds of large-scale regularities that could not be explained in terms of movements of matter or transmission of influences at speeds not greater than that of light.

What have these arguments to say about the deeper nature of time? As pointed out in Chapter 4, science has, up to now, emphasized the sequential order of successive changes. In the larger scale this includes, for example, the theory of evolution. In the generative order, however, time is not put into the first place. Rather, time has to be related in a fundamental way to the generative order. The image of a stream is helpful in this respect. The stream can be studied by following an object that floats along it, in a time process. However, it is also possible to consider the entire stream all at once, to reveal the overall generative order that goes downstream from the source or origin.

But the stream is only an image. The essential flow is not from one place to another but a movement within the implicate and superimplicate (generative) orders. At every moment, the totality of these orders is present and enfolded throughout all space so, just as with successive enfoldments of the ink droplets, they all interpenetrate. The flux or flow is therefore between different stages and developments of these orders. However, because of the possibility of loops, this flow may go in a pair of opposite "directions" at the same time.

The temporal process of evolution of the universe is constantly generated within this flow from a "source" or "origin" that is infinitely far into the implicate and generative orders. To see the universe in this way is to see "the whole stream at once" and this perception may be called *timeless*, in the sense that what is seen does not involve time in an essential way. However, the modes of generation and unfoldment in the stream imply that everything changes in successive moments of time. So in the flux described above, the timeless order and the time order enter into a fundamental relationship. However, because this relationship is now seen through the generative order, the time order appears very different from what it is in the traditional approach. It is not *primarily* a transformation within a given level of organization and explication. Rather it is, in the first place, a transformation of the entire "stream" of the implicate and generative orders that takes place from one moment to the next.

This discussion appears, at first sight, to reduce the time order so that it could, in principle, be derived completely from the timeless order. This would indeed be so, if the "flow" in the implicate, generative stream were only in the "direction" from the source or origin down to ever more explicate orders of succession. However, because of the two-way nature of this flow, there is an inherent dynamism in the theory and such a reduction is not actually possible. The timeless order and the temporal order therefore both make essential contributions to the overall order.

What is called for is to consider both of these as extremes and to explore the question of where there is not a rich new domain of order that is intermediate between the two. Until now, physics has not, however, been carried far enough to provide a context of sufficient subtlety to allow questions of this general nature to be studied usefully. Within the present chapter, this will, however, be done in other fields. But with regard to physics, so far the "stream" has been studied only up to the stage of the second implicate order. This contains the "information" represented by the wave function of the universe. This information affects the fields and particles in ways that have been described earlier so that they no longer satisfy the classical equations of motions, and the laws of quantum theory then apply. But if the information from "higher up the stream" also becomes active in this context, the laws of the quantum theory may cease to hold, for these come about by abstraction from the total stream. If this should happen, then physics could begin to provide more indications of the nature of the "stream" than it does now.

According to this, the big bang no longer has the unique significance that it does in the historical-evolutionary approach. In any case, reality contains immensely more than science may happen to know, at this moment, about the universe. For example, the universe may involve laws that go far beyond those on which the current theory of the big bang is based. Therefore it is quite possible that the big bang is only incidental in a totality that is immeasurably more than anyone could ever hope to grasp as a whole.

Current quantum field theory implies that what appears to be empty space contains an immense "zero point energy," coming from all the quantum fields that are contained in this space. Matter is then a relatively small wave or disturbance on top of this "ocean" of energy. Using reasonable assumptions, the energy of one cubic centimeter of space is far greater than would be available from the nuclear disintegration of all the matter in the known universe! Matter is therefore a "small ripple" on this ocean of energy. But since we, too, are constituted of this matter, we no more see the "ocean" than probably does a fish swimming in the ocean see the water.

What appears from our point of view to be a big bang is thus, from the perspective of the ocean, just a rather small ripple. On an actual ocean, waves from all around sometimes fortuitously combine at a certain point to produce a sudden surge of such magnitude that it can overturn a small ship. For the sake of illustration it is possible to suppose that what we call the universe originated in a similar way.

THE GENERATIVE ORDER OF LIFE AND ITS EVOLUTION

The current approach in biology and the theory of evolution is to treat explicate and sequential orders as basic. It is assumed that ultimately everything in nature can be reduced to explanations using such orders, for example, in terms of atoms, molecules, DNA, cells, and other structures. But if the generative order is basic to inanimate matter, then it is even more essential for the understanding of life. In the explicate and sequential orders, life appears to arise as a fortuitous chance combination of molecules which leads, in a more or less mechanically determined way, to further developments which produce ever higher and more complex forms. While this approach can be admitted as significant for study, it is now seen as an abstraction and approximation in the light of the generative order. Its deeper meaning is to be understood by exploring how it reveals the inward generative order of the "whole stream" that is constantly present.

In this connection some scientists, notably Erwin Schrödinger, have suggested that the quantum theory, with its new feature of wholeness, could explain the basic qualities of life. However, current work in biology hardly takes the quantum theory into account, even though it is necessary for understanding the very existence of molecules. The current approach is justified by pointing out that the relatively high temperatures at which life becomes possible make long-range quantum connections not particularly important (although it may turn out that in certain macromolecular processes, such as protein folding, long-range quantum correlations may indeed be relevant). In conclusion, therefore, it does not appear likely that the essence of life is to be understood in terms of the details of conventional quantum theory.

Rather it is necessary, for the understanding of life, to go beyond the quantum theory and the superimplicate order, into an infinity of generative and implicate orders from which present theory has been abstracted. In doing so, however, it is not intended to seek the "ultimate origin of life" in a reductionist way by going, for example, to an even more fundamental microscopic theory than the quantum theory. Rather it is being proposed that a deeper generative order is common to all life and to inanimate matter as well. It is not therefore an attempt to explain life in terms of matter, but rather to see how both emerge out of a common overall generative order. Within this order there is room for new kinds of "pools of information" from which life could be generated. The wholeness of the living being, and even more of the conscious being, can then be understood in a natural way, rather as the wholeness of the molecule and the superconducting system is understood (although it must not be forgotten that life is much subtler and more complex than molecules and superconducting systems). Life is no longer seen as the result of somewhat fortuitous factors, which perhaps happened only on an isolated planet, such as Earth. Rather it is seen to be enfolded universally, deep within the generative order.

Consider the implications of this approach for the theory of evolution. While the current neo-Darwinian theory is valid in its proper domain, it represents an abstraction from a much larger implicate and generative order, and its main significance is to be found in its relationship to this latter order. As a matter of fact, there are a number of features, within the neo-Darwinian view, that appear to be unsatisfactory and that give rise to problems of interpretation. In particular, an individual organism is regarded as being an expression of its genetic material, within a particular environment. To be sure, this environment has an effect on the individual expression of genetic material and on the variations in the overall form of a plant or animal. For example, plants of identical genetic stock but grown in the absence of certain minerals will be stunted in comparison with those grown under normal conditions. However, it is believed that the environment has no effect on the genetic material itself, other than to contribute to random mutations and determine, through natural selection, which forms will survive. Hence, while the environment affects the actual physical makeup of the individual plant or animal, its only action on the genetic material itself is to induce random changes. In this sense chance is assumed to play an absolute role in evolution as well as in the origin of life; just as it is believed to be absolute within the conventional interpretation of the quantum theory.

While this overall approach may be adequate when it comes to the explanation of relatively small changes and variations, it may not be sufficient to account for more spectacular changes, such as the appearance of an entirely new species. For major evolutionary changes require the coordinated development of many different pieces of genetic code. It is not sufficient, for example, to give a bird wings if it is to fly. In addition its bones must be made lighter while at the same time maintaining their strength, feathers must be aerodynamically adapted, the center of gravity must shift, the breastbone and musculature must develop, and changes in metabolism are required to provide sufficient energy for flight. If such changes do not all occur together and in a coordinated fashion, then they may well be disadvantageous to survival. It is difficult to understand how so many of the highly coordinated chances demanded by evolution could have come about by fortuitous chance combinations of small random mutations.

It is proposed here that the generative order enfolds orders that go beyond those that can be contained even in the quantum domain, orders that have so far been "hidden" to science. Such orders may involve implicit totalities of structure of the organism which can all emerge together. This is somewhat similar to Goethe's notion of the "original plant." Goethe was speaking of, not a beginning in time, but rather a source of order in the generative principle that implicitly contains a totality of structures and forms of a range of related species. Orders of the kind discussed above may be called a kind of *"protointelligence."*[1] This does not refer to a form of conscious awareness but to the active generation of new order, from levels that are at present "hidden" to science, rather than to a more or less mechanical adaptation to an existing order.

1. The astrophysicist Fred Hoyle has also argued that life on Earth involves the action of some form of intelligence which pervades the entire universe. His ideas are given in *The Intelligent Universe*, Michael Joseph, London, 1983.

This protointelligence could be regarded as manifesting through a kind of "free play" similar to that in which, it is suggested, actual human intelligence emerges in thought and action. Under ordinary conditions, the scope of this free play is limited, as in the neo-Darwinian approach, to small variations of a more or less random nature. However, it seems to be generally accepted that when species undergo large mutations, for example, from reptiles to birds, the process is so rapid that it gives rise to few intermediate fossil records, if any. To explain this, it is proposed that, for reasons not yet known, the genetic material of organisms can, from time to time, enter into a state of greater scope for "free play" in which whole new totalities of structure may be put forth, as it were, for a "test" of their viability.

The essential feature of the protointelligence would be that these totalities are not formed in a random fashion, but that they emerge as relatively integrated wholes from subtler levels that are enfolded beyond the first and second implicate orders. (These latter orders, as has been pointed out, are all that physics and chemistry are currently able to treat.) If such new structures continue to survive for some time then, as happens with thoughts that survive sustained tests of validity, the structures will become stabler. After this they will undergo only small random variations until, once again, conditions arise in which the genetic material becomes capable of a wider range of free play.

But what determines the conditions under which this "free play" operates? One possible answer is to consider, along the lines discussed in Chapter 3, that each part of the universe is a kind of analogy to the whole and is therefore structurally related to the whole. In this way there could be a kind of subtle perceptive response to the environment which would, under appropriate circumstances, give rise to new kinds of active information. This would dispose the organisms to enter a state of "free play," in which new totalities of structure are "proposed" which are in some kind of harmonious "proportions" or "ratios" with the environment. Perhaps there could even be a closed loop, similar to that within the computer game, in which this active information is able to feed back into the "perception." This whole process would, of course, take place within the general context of natural selection, which would act as a "test" of these "proposals."

A number of biologists, including the late C. H. Waddington, Stephen Jay Gould, Brian Goodwin, and Rupert Sheldrake, have expressed serious reservations with regard to the current approach to the whole question of morphology, especially in connection with the appearance of new species.[2] In particular, Sheldrake has suggested that the accumulated experiences of a species constitute a kind of formative memory, which disposes individual members to develop accordingly. From the present point of view, this could be regarded as a relative fixing of the primary creative action of the protointelligence in matter. This relatively fixed form would therefore be closer to a proto*intellect*.

The approach now being proposed contains something of Lamarckism and Darwinism while going beyond them. The nature of protointelligence has evident similarities to Lamarckism in the sense that it creatively responds to a particular situation. On the other hand, the notion of protointellect is closer to Darwin, in the sense that the species depends on "memories" contained in the DNA, and possibly in the formative fields of Sheldrake. However, this is only part of a total generative order that goes on ultimately to include all life, all matter, the whole known universe, and what may lie beyond.

THE GENERATIVE ORDER IN SOCIETY

This discussion of the generative order in physics, cosmology, life, and evolution can now also be applied to human societies, where similar ideas have already been hinted at. For example, Arnold Toynbee proposed that new civilizations arise out of a creative surge, expressed most strongly perhaps in

2. The ideas of Waddington and others are to be found in the three-volume report on a particularly interesting symposium held in Bellaggio, Italy: C. H. Waddington, ed., *Towards a Theoretical Biology*, Edinburgh University Press, Edinburgh, and Aldine Press, Chicago, 1969. For Sheldrake's ideas, see Rupert Sheldrake, *A New Science of Life*, Tarcher, Los Angeles, 1982.

certain individuals.[3] Following this surge these civilizations flower, are maintained for a time, but eventually decay. The notion that Western civilization is subject to such a cycle is not uncommon. The Renaissance, in which our present civilization has its roots, may be considered as the creative surge that was partly inspired by the example of the creative surge of the ancient Greeks. But as early as the eighteenth century, Diderot, in *Rameau's Nephew*, saw a fading and decay of this impulse, and later writers, such as Nietzsche and Spengler, spoke openly on the death of God and the decline of the West. In modern times, this theme of decay has become a major one, not only in literature and drama but, at least implicitly, in a considerable part of the visual arts. This suggests that the generative order of societies has two basic sides, that of growth and that of decay.

The most stable society known is that of the Australian aborigines, which anthropologists estimate has lasted for about thirty thousand years in more or less its present form. The ancient Egyptian society lasted for several thousand years, but with a number of ups and downs. The Chinese and Indian societies have roughly a similar history. The ancient Greek and Roman societies were much less enduring. Modern society has, thus far, existed for a relatively short time, and it is now common to predict that it may not last much longer.

The instability of societies is, of course, partly due to external invasions, but history has shown that a stronger cause can generally be found in internal conflict and breakdown. The Australian aboriginal society does not appear to have had conflicts serious enough to cause internal decay. However, that society was not only relatively simple but was, within certain bounds, fairly fluid and mobile in its structure, so there was no elaborate organization that could easily break down. Moreover, at least in relation to nature, there was a pervasive creativity in daily life. This society was therefore able to go on meeting the rather simple challenges it had to face for tens of thousands of

3. Arnold Toynbee, *A Study of History*, Oxford University Press, New York, 1947.

years. With the coming of the Europeans, the very simplicity of this society became a serious limit, in the sense that it lacked the degree of subtlety and complexity that would have been needed to meet this challenge. By contrast, the Chinese society showed its ability to absorb its invaders and make them Chinese.

Clearly, the notion of mere survival is much too simple for discussing the evolution of societies; rather it is the ability to make dynamic and creative responses to whatever changes may take place. A society that has gone beyond its first creative surge hangs on indefinitely to the habitual orders that are contained in its customs, taboos, laws, and rules and that are held in its unconscious infrastructure. Because these orders are fixed and limited, they will be bound, eventually, to become inadequate, in the face of the ever-changing nature of reality. Changes may be precipitated by natural environmental occurrences, such as the failure of crops or game, or by the pressure of external groups. But more generally, especially as civilization grows more complex in its structure, what is most significant is the activities of society itself. These may lead eventually to decay, more or less independently of the institutions, will, and desires of the people who make up this society.

In nature as a whole, orders of growth and decay are inseparably interwoven, as two sides of one overall generative process. For example, with plants such growth and decay give rise to the ground out of which can grow other plants. Indeed, without the death of the individual organism, life would not go on. It is also a rather common feeling that perhaps it is natural for societies to take part in this cycle, so that the ultimate decay of each society is therefore inevitable. In early times, such as the Stone Age, this cycle was, in a certain sense, a viable possibility for humanity as a whole. Decay could take place in some part of the world, while growth and flowering occurred in other parts. Furthermore the decay of one society could provide a kind of fertile ground for the growth of a new one. Thus it can be plausibly argued that this whole cycle is in some sense conducive to creativity.

But even if this argument is accepted as applying in the

past, it should now be clear that the very forces that have been liberated by the development of modern societies make this kind of process no longer appropriate. What faces humanity today is not the continuing cycle of civilization but worldwide annihilation, which may even destroy the order of nature on which all life depends. If the cycle of birth and decay continues, it would take a very optimistic person indeed to say that the human race will survive for as much as a thousand years, which is after all a rather short period in human history, and much less in human evolution.

The challenge that faces humanity is unique, for it has never occurred before. Clearly a new kind of creative surge is needed to meet it. This has to include not just a new way of doing science but a new approach to society, and even more, a new kind of consciousness.

A consideration of the historical background of this state of affairs may be helpful in understanding what is required for such a transformation. Arnold Toynbee has described the origin of ancient Greek society as lying in a creative impulse which emerged out of its response to some kind of challenge that it faced in those early times. It is difficult at such a historical distance to analyze this response in detail. However, Greek art, in its early vigorous phases, certainly reflects a creative energy that must have characterized the society as a whole. This energy led to a flowering not only in art but also in philosophy, mathematics, theater, poetry, and the development of democracy in politics. But within a relatively short time this had begun to decay, partly as a result of internal dissension. In art it is possible to see this process of losing touch with the creative surge as works become elaborately refined and eventually cease to express the original vital energy.

In the Renaissance, there appeared a whole new spirit of the age, which could be characterized as "the new secular order." In the previous eternal order, the highest value was that which never changed. But in the Renaissance the way things changed and the power to change them became the highest value. For example, Galileo argued that it is precisely the transformations of nature that are the most interesting. Francis Bacon carried this further, opening up the prospect that through the understanding of such processes humanity could dominate nature for the general good. In sculpture Michelangelo explored the dynamic transformations of human form that could be contained within a single block of marble, and Leonardo, in his notebooks, was preoccupied with energy, change, the Deluge, and the general expression of power in nature. Throughout the whole culture, there spread a wave of infectious exhilaration. This released the energy necessary for a vast surge of new creative activity. Clearly, a new vision had entered the generative order of European society.

If all societies have similarly started from a creative perception in the generative order, then how do those of appreciable complexity begin to decay? A basic reason may be that as new knowledge and institutional structures accumulate, they tend to become ever more rigidly fixed. Eventually it seems as if the whole order of society, and the security of each individual, would be endangered if these structures were to be seriously questioned or changed. In order to hold things fixed, therefore, the culture feels itself forced to react in ways that are basically false, for example, by becoming fragmented. In Chapter 1 it was shown that by ignoring the connections of each thing to its whole context, the illusion can be created that the ideas, structures, and institutions that are the dearest can go on indefinitely and unchanged. This fragmentation is itself part of a broader process in which society and culture "defend themselves" against acknowledging the actual facts that call for change. Thus, as pointed out in Chapter 1, the mind is then caught up in the general process of "playing false" in order to create illusion and delusion. This leads society into collusions in order to defend the illusions and delusions. This attitude has also been called "bad faith."

Clearly this process is a breaking down of the basic generative order of the society. Those who see this will call for a fundamental and revolutionary change. But this provokes only fear and hatred because so many people believe that society cannot function in any other way and, for many other reasons as well, do not wish for change. Even in those cases where change is able to take place, it is limited by the very factors that lie behind the original decay in the generative order in the first place. People now become disillusioned by the ultimate and deep failure of their efforts to change. They are lost in despair, or else they entertain false hopes which are essentially based on fantasy. This leads to a further stage in the breaking up of the generative order in which those in authority attempt to establish more and more rigid control in order to prevent further deterioration. But rigidity is ultimately the very source of this deterioration, so things can only get worse in the long run. Indeed the whole process of breakdown is sustained and constantly extended, because all the proposed remedies are actually different forms of the same illness that they aim to cure.

The breakdown in the generative order of society is more than mere decay, which might be regarded as the opposite side to a natural process of growth, in the generative order of society. Rather, the rigidity in the generative order constitutes an extremely pervasive and far-reaching blockage of free play of the mind, and this makes for the constant spread of false play and prevents creativity that could adequately meet this situation. Indeed, as with the individual, to the extent to which society is no longer creative in its basic generative order, it becomes destructive to itself and to everything that it touches. Thus both with regard to human beings and society as a whole, the basic distinction in the generative order is not between growth and decay, but rather between creativity and destructiveness.

Becoming caught up in the false processes of a destructive generative order, as described above, has been the ultimate fate of all known civilizations. But is this inevitable? If so, the outlook for the human race is very grim indeed. For the means of destruction are now global and almost total, and no part can be isolated from the rest of the world.

Need this cataclysmic failure of the human race necessarily take place? Can it be avoided by a creative approach to all areas of life? What is needed is not simply a new creative surge but a *new order* of creative surge. One that extends into science, culture, social organization, and consciousness itself. In order to explore this issue properly, it will first be necessary to enquire into the nature of consciousness, for this contains the key to both creativity and the factors that are impeding it.

CONSCIOUSNESS AND MATTER

There is a great deal of evidence that consciousness is inseparable from the material processes of the brain and nervous system, and indeed of the whole body. Thought, feeling, and intention can radically change the distribution of blood flow and various chemicals within the brain. Likewise, changes induced in the brain's chemistry can modify overall mental states. More subtly, abstract information concerning the meaning of some external circumstance, such as the presence of danger, can affect the hormones, such as adrenaline as well as the neurotransmitters, neuropeptides, and other brain chemicals, when they move out through the whole body. In this fashion, brain and body, in turn, profoundly change thought, feelings, and intention.

One commonly accepted view of all this is that ultimately the whole of mental function can be reduced to nothing more than physical and chemical processes of this kind. Thus it is held by many neuroscientists that "consciousness is an epiphenomenon of the brain." An opposing view is that consciousness is a pure disembodied spirit in some kind of interaction with matter.

Rather than arguing the merits and drawbacks of these positions, an alternative view will be advanced. This is based on the notion that reality is inexhaustible and whatever we say a thing is, it is something more and also something different. Hence, for example, if we say that consciousness is a material process, this may well be fairly accurate up to a point. But it is also more. Its ground is in the infinite depths of the implicate and generative orders, going from the relatively manifest on to ever greater subtlety.

The word *subtle* means elusive, rarefied, delicate, intangible, and ultimately indefinable. This would at first sight suggest that consciousness is some form of disembodied spirit. However, in view of the neurophysiological evidence cited above, which has of course only relatively recently been made available, it seems reasonable to conclude that consciousness always has a material aspect or side that can in principle be studied scientifically. But in this latter endeavor, no matter how far science goes, there is always more and the totality will always elude the grasp of science. That elusive side, which is nevertheless always present in some form, is generally called the mental aspect. Consciousness can then be understood, at each stage, as the interweaving of these two sides. Hence it is possible to go beyond the usual approaches in which mind and matter are two separate but interacting streams, or in which consciousness is considered as just a material process.

This way of looking at the nature of reality can be extended to all life, and even to what is normally called inanimate matter. The root of what is manifest in these forms lies in the ultimate depths of the implicate and generative orders of the totality of matter, life, and mind. In this sense, therefore, even inanimate matter must have some kind of mental aspect. Indeed it has already been shown that the first implicate order in quantum mechanics is organized by information in the second, superimplicate order. Therefore, for the first implicate order, such information could be regarded as a subtler mental side. Likewise, for the second implicate order, the third is a more subtle mental side and so on. Of course, this does not imply that "consciousness" can be imputed to electrons or to other such "particles." This arises only at much deeper levels of the generative order.

The essential point, however, is that there is no absolutely sharp "cut" or break between consciousness, life, and matter, whether animate or inanimate. Of course each of these can be analyzed in thought as categories with a degree of relative independence upon each other. This makes it possible for each to be studied, up to a point, in its own right. But more generally, each of the stages of consciousness described above can be regarded as lying somewhere along the "stream" of the generative order. As its origin is approached, the subtle mental side becomes more and more important, while in the journey "downstream," the manifest physical side becomes more evident.

AWARENESS AND ATTENTION

If consciousness always has, at each stage, a subtler mental side and a more manifest material side, the question then naturally arises: How are these two sides related? It will be argued that this relationship comes about, in the first instance, through awareness and attention.

In order to see what is meant by awareness, note that according to its Latin root, *consciousness* is "what is known all together." Originally this meant "what everybody knows all together" and thus referred to something that is essentially social and cultural. By now, however, it generally refers to "what the individual knows all together," that is, to the total state of "knowingness" of the individual. This change of meaning goes together with the change in the order of society, which was discussed earlier, in which the individual came to have an ever-increasing prominence.

If consciousness is some kind of "knowingness," then what is meant by the "unconscious mind"? Is it an "unknowing knowingness"? The resolution of this paradox depends on bringing in the distinction between consciousness and awareness. These terms are generally used in a fairly interchangeable way and yet they do have different connotations. Awareness is based on the word *wary* or *aware*, meaning "watchfulness" or "heedfulness." The term *sensitive awareness* suggests the image of a person who is very watchful and perceptive and therefore disposed to respond even to the subtlest impressions of all kinds. Such watchfulness may, for example, be precipitated by the presence of danger. This sensitivity is not, however, primarily concerned with already organized knowledge. Rather it responds to subtle differences, similarities, and relationships in impressions from sense organs, muscular movements, reactions, feelings, and thoughts, and in "ratios" of all kinds, both material and mental. This sensitivity is the source of all information that may later give rise to a perception and knowledge of form, order, structure, and ultimately all that has meaning in consciousness. It is constantly transforming any given context to an ever subtler level, and thus plays an essential part in relating mental and material sides at each stage of consciousness.

Evidently, it is possible to have consciousness, that is, "knowingness," with little or no awareness. For example, a great many of our habitual and automatic responses imply a considerable amount of knowledge. This includes not only abstract knowledge but concrete knowledge in the form of skills that are needed to deal with familiar situations. Yet the individual who carries out these responses is often hardly aware of what he or she is doing. That is to say, there is little sensitivity to most of the differences, similarities, and relationships that could give rise to new responses. Instead the individual operates from largely "unconscious" physical and mental predispositions.

The term *conscious awareness* is in fairly common usage, and is taken to mean a consciousness (i.e., a knowingness) that is pervaded, to one degree or another, with a sensitivity to the immediate processes of environment, body, and mind. Rather than using the self-contradictory term *the unconscious*, meaning by this "unconscious consciousness," it would seem better to distinguish between unaware consciousness and aware consciousness, or alternatively, conscious awareness.

A very large part of what may properly be termed *consciousness* or *knowingness* is not normally accessible to awareness, and indeed, there is no reason why it should be accessible. This includes the kinds of "knowledge" that are built into the structure of the brain and body: for example, the activity of the autonomic nervous system, the processes determining motor control, and a whole range of other functions, such as the release of various hormones and neurochemicals. Also there are areas closer to awareness, for example, tacit knowledge, such as that involved in riding a bicycle, to which the philosopher Polani refers. And in addition, there is the action deep within the generative order from which creativity emerges.

Such areas are not usually accessible to awareness and may

be generally regarded as being in various kinds of implicate orders. However, it is possible, for example, in listening to music, to become directly aware of some of these implicate orders. Indeed it would seem that there is no absolutely sharp boundary between the kind of content that can enter awareness and the kind that cannot. It is possible to become sensitive to just how the muscles are affected by impulses to act and thus become somewhat aware of the tacit knowledge that is involved in riding a bicycle. Nevertheless, although the boundary of awareness may be moved, perhaps even a long way, conscious awareness is, broadly speaking, only a small part of the total field of consciousness or knowingness.

Another part of consciousness that remains fairly unaware is the tacit infrastructure of ideas, which has been discussed throughout this book. When this structure operates properly, it serves us well by providing "knowledge at our fingertips." In this way our minds are then free to concern themselves with other things that require conscious awareness. However, as happens in science, this infrastructure can also become so rigid that it interferes with proper awareness. Scientists then take a permanent bias, for example, in favor of whatever gives rise to pleasant sensations which imply harmony and security. Equally they react to sensations of disturbance or insecurity. Such a permanent bias in sensitivity is evidently not compatible with proper awareness. Rather it could better be called a kind of *unawareness*, in which the response of the mind is inhibited or blotted out in certain directions, while it is exaggerated in others. In this way, the mind is caught up in playing false with sensitivity and awareness.

Any discussion of awareness must, however, bring in the question of *attention*, which is closely related. Indeed, the two words are to some extent interchangeable, insofar as awareness can mean heedfulness, which also signifies attentiveness. Nevertheless, there is an important difference of connotation between these two words. Thus, the word *attention* means literally "stretching the mind toward something." This implies an inner activity that is needed to grasp the object of interest mentally. Even physically it has been demonstrated that a kind of attention must operate as, for example, in the eye's scanning pattern, which varies according to each person's concept of the object that is being looked at. With an inappropriate scanning pattern, this would mean that the object could not be seen properly. In a similar way, suppose that the mind is able to "scan" its overall content, including not only knowledge, but also what is revealed in the sensitive response of awareness. The mind is thus able to "take hold" of this content and grasp it as a kind of whole, at a higher and subtler level. This is the beginning of attention within the mind, which can move, if necessary, to indefinitely subtler levels. This movement, along with the sensitivity of awareness, is constantly bringing content, on the more physical side, toward the subtle "mental" side (which is closer to the origin of the generative order).

As happens with awareness, however, rigidities in the tacit infrastructure of consciousness act to interfere with the sort of free movement of the mind that is needed for proper attention. In this connection it is important to recall that each object and each aspect of the content of consciousness has to be "scanned" in a way that is appropriate to it. Rigid assumptions as to the nature of these objects or aspects lead to an inappropriate or limited scanning pattern and to the inability to see properly in unfamiliar contexts. When combined with permanent biases in sensitivity and with a variety of diversionary and distracting tactics, such a mental habit produces sudden jolts, leaps of excitement, or absorbing but irrelevant thoughts. These capture and beguile the attention and keep it in an area that is judged not to be too disturbing. In this way the mind is capable of grasping only a restricted content, determined largely by the tacit infrastructure of consciousness, for whatever content is outside the range of this attention is not seen. This whole response therefore involves being caught up in playing false with attention, which can be regarded as a kind of *inattention*.

Of course, both with attention and awareness, it is often necessary to focus for some time on a given context. For example, if a person is thinking mainly about accomplishing a certain task, then it is important to concentrate upon the proper order that is needed for carrying out the task. But if the mind holds on to this order in an extremely rigid manner, then creative new perceptions will not be able to emerge. For creativity to flower there must be an easy mobility of sensitive awareness so that attention can move freely and in any order that may be relevant at the moment in question. Clearly the free movement of awareness and attention has no inherent restrictions and is limited only by the genuine necessities of the moment and by permanently rigid features of the tacit infrastructure of consciousness. Such free movement of awareness and attention is closely related to the free play of thought, which was discussed in earlier chapters. Indeed, both these kinds of freedom, which are ultimately one, are necessary for creativity.

CREATIVE INTELLIGENCE

Awareness and attention bring about a movement of content from the more manifest physical levels toward the subtler levels of the generative order. The response to this is a movement in the other direction, an unfoldment of the creative action of intelligence. This originates, ultimately, in the depths of the generative order.

In earlier chapters it was shown how, in the free play of thought, creative intelligence responds to opposition and contradiction with new proposals. These are "put forth" for testing, in further thought and action. A similar response also takes place in the area of sense perception. When something new is encountered, which does not fit into what is already known, creative intelligence can put forth new sensory orders and structures that form into new perceptions. These are tested in cycles of perception-action, which were described in Chapters 1 and 2. Such a perceptive response to creative intelligence may, however, be not only sensate but also aesthetic, kinesthetic, and emotional. It can in principle happen in any area of life, but is especially evident in music and dance, in which a creative orchestration of themes and movements unfolds from and enfolds into a similar orchestration of very subtle feelings and impulses to act.

All of this emphasizes the universality of creative intelligence, which has ultimately the same origin in every area of life. In this connection it is not appropriate to think of experience as being something which exists on its own, and which is from time to time somewhat modified by the perceptions, thoughts, and actions that come out of creative intelligence. Rather, every aspect of such experience, whether physical or mental, emotional or intellectual, can be profoundly affected by creative intelligence, wherever this is able to act. For through its action everything may take on a new meaning.

Not only is intelligence ultimately indefinable in its origin, in the depths of the generative order, but it is also intrinsically whole. For certain limited purposes, it may of course be abstracted from the total response of intelligence and be treated as if it were a part of life that had a definable source. But basically, because of the extreme subtlety and pervasiveness of the action of intelligence, such abstractions are only of limited validity. For intelligence cannot be separated from the whole and assigned to any defined structure or order. Thus, no matter how far the generative order is investigated in thought, there is always an unlimited horizon beyond, which is too subtle to be considered in the level in which this investigation was stopped. Probably even the notion of the generative order itself is not adequate for going to the ultimate origin of intelligence, which will always be more than we can say and different from it. This also means that intelligence cannot be properly understood as something that has evolved in the order of time. Rather, for the proper discussion of the origin of intelligence, it is necessary to bring in the timeless order (discussed earlier) in a primary role, while time itself plays a relatively secondary part, in the description of the *manifestation* of intelligence.

In this respect, a large part of what is commonly called intelligence should properly be called *intellect* (this was pointed out in Chapter 3), which comprises relatively fixed knowledge and skills of various kinds. The intellect is essentially based on the tacit infrastructure of consciousness and functions with little conscious awareness. It is a little like a computer program, although it goes far beyond anything that a computer could accomplish today.

It should be emphasized in this connection that intellect, emotion, and will cannot actually be separated, except for the purpose of analysis in thought. For example, the categories of the intellect can have profound emotional impact. Words expressing totality, such as *all, always, forever, never,* and *only,* are the key ingredients of many popular songs, which weave these words together in a context whose meaning is aimed at stirring up all sorts of strong feelings. Demagogues also use such words with telling effect. Yet these totality words are in fact the most intellectual part of the language. Vice versa, powerful emotions very strongly affect the course of thought; indeed without some emotional arousal, we would think very little at all. In the case of the will, it is hardly necessary to point out that it depends crucially on the content of thought, without which it would not have the necessary *determination*. This content depends in turn on intellectual concepts, such as necessity, the root of which is *ne-cess,* meaning "not yielding." Necessity, therefore, becomes a disposition not to be readily deflected, which is, of course, an essential characteristic of the will. In short, it is not possible to determine where the will begins and intellect ends.

The inseparable nature of emotion, intellect, and will is in harmony with what is known about the general structure of the brain. For example, a very thick bundle of nerves connects the frontal lobes, which have an intellectual function, to the base of the brain, which is more associated with the emotions and from which the whole organism is bathed in chemicals that affect all parts profoundly. Recent knowledge goes much further toward revealing a similar but far more extensive and pervasive interconnection in various structures and processes that must be involved in the actual operation of thought, feeling, and will. The activity of each individual nerve cell is directly connected, via synaptic links which involve the activity of various neurochemicals, to some thousand other nerves. In any neural network, therefore, the number of interconnections is astronomical. In turn, the strength of each interconnection is influenced by neurochemicals, such as the neurotransmitters, as well as by the actual electrical activity within the network. The system is of an almost unanalyzable complexity and subtlety in the operation of its various processes, for individual nerve impulses are mediated by an enormous range of chemical and electrical responses, some of them local and others global, some general and others highly specific in nature. An extreme reductionist view may suggest that the nerve impulses are "processing data" relating to movement, senses, and the intellect, while the neurochemical bath would be close to a basis of experiencing an emotional response. But in view of the enormous complexity of the brain, such an image is clearly too crude a simplification. Rather, the insights of neurochemistry and the nature of nerve networks indicate very strongly that there can be no fundamental separation at this level between thought, feeling, and will.

The subtle mental side to these processes indicates that there is even less reason for making such a separation. For it is possible to sense and experience directly an intimate connection between thought, feeling, and will and show that there is no point at which one of them ends and the others begin. Moreover, creative intelligence can profoundly affect the whole meaning of these functions, as well as the entire way in which they proceed at the level of physics and chemistry. Indeed, it has already been suggested why a large part of this content probably cannot be understood in terms of the current laws of physics and chemistry and requires a level of explanation that goes beyond the superimplicate order.

Ultimately the origin of all of this lies in the creative intelligence, which is beyond anything that can be discussed in the manifest physical side. This intelligence is universal and acts in every area of mental operation.

In its depths, such intelligence can involve no separation between knowing, feeling, and will. Thus, one way in which intelligence becomes manifest is by organizing the categories, orders, and structures of the intellect in new ways. It may orchestrate feelings in an ever-changing movement, like that which is experienced in music. Such a movement goes beyond the sort of succession of fixed patterns of feelings that can be identified in fairly well defined forms, such as pain, pleasure, fear, anger, desire, and hatred. Orchestrated movement of feelings may perhaps be what is meant most deeply by words such as *love, beauty, vitality,* and so on. But where these feelings emerge from the whole of the generative order, they must evidently have meanings that are not adequately signified by what is commonly conveyed by these words.

In everyday consciousness, however, the mind is absorbed largely in the tacit infrastructure of ideas and dispositions to feel and act, which are mainly mechanical in their operation. In a metaphorical sense, at least, this activity of the mind could be said to be "programmed." But it should be clear that these programs, while both useful and necessary, are limited, since something more and something different, creative intelligence, is always in principle available.

THE BRAIN AND ARTIFICIAL INTELLIGENCE

Within the current, generally mechanistic and reductionist, worldview of science, it is only natural that a serious attempt should be made to model the function of the brain on some extension of what is now meant by a computer. This has led to the attempt to develop what is called artificial intelligence, a field that is associated with several different approaches, some of which are quite subtle. For example, the field of cognitive psychology includes the study of perception, language, problem solving, and concept formation. Other research fields combine with the neurosciences, particularly in the areas of vision and motor control. Other lines of investigation are leading scientists to develop computers that are far more sophisticated than those in use today. For example, there are "massively parallel" computers, analog systems, and computer languages that deal with concepts directly and are able to examine their own strategies and goals.[4] Such research has great value, but

4. Further discussion of such developments is given in F. David Peat's *Artificial Intelligence: How Machines Think*, Baen Books, New York, 1985.

much of it seems to be based on postulates that are of a generally reductionist nature. One of these is that all cognitive processes can ultimately be revealed through sufficiently subtle experiments and from inferences that are drawn from these investigations. Another assumption, perhaps more fundamental, is that at some stage scientists will come to the end of this cognitive analysis, or if not, then what lies beyond will not be particularly significant.

It should be clear that these assumptions are not in harmony with the approach taken in this book, where cognitive processes are based on the intellect alone. Clearly this does not capture the essence of the whole generative order of the mind. Very probably it will be possible to simulate an unlimited number of aspects of the intellect, which is after all a relatively mechanical crystallization of the intelligence. In this sense, the proper description of these studies should be called *artificial intellect*.

However, if scientists still claim that artificial *intelligence* and not artificial *intellect* is genuinely involved in their researches, then certain important questions must be answered, some of which are indeed already being considered by researchers in this field. The basic problem is similar to that which is also encountered in human consciousness: How is it possible to question fundamental assumptions that have been fixed as necessary for the whole system of operations within the computer? In the case of a computer, programs and strategies correspond to fixed tacit assumptions in the mind. But these assumptions are just what must change in an appropriate way to meet the challenge of new and hitherto unknown situations. One solution is to have additional "higher order" internal or external computers that control these changes. But these machines would themselves also contain fixed elements in their programs that would in turn be required to change.

Moreover, the design of these fixed elements, as well as indeed that of the whole system, depends on a host of further assumptions made by the designers, most of which are part of the tacit infrastructure of the general consciousness of humanity. For example, in planning how to deal with a given problem, it is generally taken for granted, without any explicit discussion, that the current laws of physics and chemistry will be valid in any situation that may come up. To provide an adequate simulation of human intelligence, the computer would have to be able to become "aware" of such implicit assumptions, and to question them when necessary. This, however, would require that the computer be able to uncover and question the whole tacit and largely "unconscious" basis of the society in which it was made.

To the extent that human intelligence is able to confront such questions adequately, there is free play in thought and a corresponding free movement in awareness and attention, which makes possible the action of creative intelligence. How could this be simulated by a computer? One suggestion is to introduce random changes in some of its basic programs and strategies. However, as with Darwinian evolution, such "mutations" would for the most part be inappropriate and destructive. A possible way to deal with this would be to suggest that only those programs or strategies should survive that best fit the environment. But here, the criterion or selection has simply been externalized to include, for example, the scientists themselves, or else the human society with its requirements. If the computer itself is justifiably to be called intelligent, then it would have to contain its own internal criteria for what is beneficial or "fitting." As far as can now be seen, this again would have to be preprogrammed, and in facing a new situation it is just these very criteria which themselves may have to change. This recalls Kuhn's notion of a scientific revolution in which the criteria for judging theories are themselves subject to change.

Moreover, it can be called into question whether random mutations, of the kind discussed here, can ever give rise to totally "creative" perceptions, of the kind experienced by Bach and Mozart. This question is related to an earlier one of whether large changes in animal species can be accounted for by random mutation in DNA. In that connection it was suggested that major changes may have their origin in "hidden" orders, beyond those that are now studied in terms of current physics and chemistry. More generally it was proposed that creative intelligence originates in the subtle depths of the generative order, beyond anything that could be specified in terms of well-defined concepts at all. It is there, too, that the free play of thought and the free movement of awareness and attention have their ultimate ground. It does not seem possible for any novel design of computer or language to simulate such freedom adequately, although it may be capable of providing significant developments within the relatively limited area of artificial intellect. In conclusion, it should be clear that creative intelligence cannot be grasped by the intellect in any form and that it will necessarily elude all such attempts to capture it in this way.

CREATIVE INTELLIGENCE, TIME, AND THE TIMELESS ORDER

If creative intelligence originates in the infinitely subtle depths of the generative order, which is basically not in the order of time, then it follows that the discussion of creative intelligence must bring in this timeless order in a fundamental way. This order must be considered all at once, rather than in an order of succession. However, to do this properly requires not only a comprehension through the intellect but also a more immediate and direct perceptual contact in which there is actually a sensitive awareness and an alert attention to this "whole stream." In general this is not easy to do. We have become habituated to a limited sensitivity and an attention that are appropriate only for apprehending partial aspects of reality and for focusing on the orders of succession that are appropriate to our notion of time. This has come about through a long historical process, in which the order of time has assumed an ever-increasing importance to society in general.

In preindustrial societies, time, as an order of succession, is not strongly emphasized. Time is generally understood through the natural cycles of light and dark seasons, growth, and decay, which are in essence, aspects of an eternal order. The past is often seen timelessly through mythology, while the noncyclical aspects of time are contained largely in vague notions, such as "tomorrow," that may mean "any time in the future, perhaps never." But in a modern industrial society, the significance of time is all-pervasive. Every moment from awakening to falling asleep is ordered and organized in terms of time. The past is ordered through records, which are now greatly extended with the aid of computers, and the future is generally regarded as being fairly well mapped out for many years ahead. This notion is not cyclical but progressive. Time is considered to arise in irreversible change, aimed at various ends and goals. A great deal of pride is taken in the organization of time, and to follow time is to be regarded as virtuous. People should not "waste time," for "time is money" and punctuality is a sure sign of inner discipline and order. It is really striking to reflect on how much this attitude has been strengthened, even in the past few centuries.

It is therefore not surprising to discover that, by now, our awareness appears to operate almost solely in terms of time and that attention is almost entirely dominated by the need to "scan" an experience in an order that is appropriate to time. As a result each of us may really feel that "time is all there is," and that the notion of a timeless order is either an illusion or an empty construction of the intellect. In this way attention has been captured within a rather limited and rigid pattern.

Nevertheless there are no inherent structural limitations on the free movement of attention. It is possible for attention to turn toward the implicate and generative orders, as in the example in Chapter 4 of listening to music. Attention may operate in a similar way while contemplating nature, for example, in observing the flow of a stream. Sooner or later the overall flow of the "whole stream" is sensed, in which various subsidiary movements, such as eddies and ripples, take their places as minor variations. In such a situation, consciousness itself seems to be flowing, in a manner that is analogous to the flow of the stream of water itself. But this is not a flow in the ordinary visual or tactile space, from one place to another. Rather it is felt to be a kind of space of the mind, that is "everywhere and nowhere." It may be experienced as the implicate order, which is also generative in the sense that content from the "origin" of the flow tends to unfold farther downstream. In this implicate order, time is of secondary importance and is not very relevant to the "whole stream."

Generally speaking, as soon as one returns to the ordinary concerns of the day, this mode of perception will vanish as attention is given to patterns which emphasize time and analysis into separate but interacting parts. However, if the significance of what is happening is actually realized, then the free movement of attention may itself turn toward that which is constantly diverting attention toward time and analysis. In this way the free mode of attention may then be available whenever there is a quiet period.

It is useful to visualize the analogy of an obstruction which causes water to back up, sometimes a long way toward the origin of a stream, and in this way profoundly modifies its flow downstream. Rigidly fixed ideas similarly obstruct the flowing of the "stream" in the space of the mind. To free this flow therefore requires the removal of the obstruction. But how can this take place? The suggestion here is that if such a free flow comes about in any area, such as the observation of nature, then through proper attention to the whole question, a similar flowing movement can, as it were by analogy, come about in other areas and will act to remove the obstructions that operate there.

As attention moves toward the encompassing order, which is sensed as a kind of "present moment" that is constantly flowing, the aspect of change plays a smaller and smaller role in the "space" of such a perception. The "timeless" flow, which is constantly renewed from the origin of the generative order, is thus sensed as eternal. This is the deeper meaning of the eternal order, which was discussed in Chapter 3. That is to say, the eternal order is not properly to be regarded as static, but rather as eternally fresh and new. As attention goes to the consideration of succession, however, it begins to be directed toward the temporal or secular order.

In terms of the superimplicate order, it is clear that if the flow were only from the subtler to the more manifest, then it would reduce to a purely timeless order as described above. Such an order could in a certain sense be intensely creative. But if what happens in one moment would not be related to the next moment, such creativity would resemble an arbitrary series of kaleidoscopic changes with little total meaning. Moreover, the more manifest levels would have no autonomy in relation to the subtler ones. A more meaningful kind of creativity can be obtained by relating the eternal order to the time order, and by allowing the more manifest orders to have some degree of relative autonomy.

In the superimplicate order a similar relative autonomy of the manifest, or explicate, forms has already been explained as arising in closed loops, similar to that found in the computer game. This also introduces a relatively self-determined order of development in time. The "tighter" this loop is, the more nearly self-determined will be the order and the more nearly "secular" it will become. As the connections in the loop are "loosened," however, allowing for some degree of free play, it becomes possible for the creative action from the subtler levels to enter into the activity within such a loop. So, there is in general a whole spectrum of orders available between the eternal and the secular extremes. Within this spectrum there is a broad and rich area for human creativity. Thus, for example, the "tightness" or "looseness" of the closed loop has to vary to fit each new situation appropriately. Each change in these qualities will bring about a different order in which the timeless and temporal sides will have different degrees of importance.

Clearly an attempt to stay rigidly at one of these extremes will seriously impede creativity. For example, if the mind is rigidly set on accomplishing a given goal in a certain order of time, creativity is effectively blocked. Rather whoever has the interest and passion needed for creativity will often be hardly aware of the passage of time. However, if all awareness of time order were permanently blotted out, then a person would be incapable of the consecutive, sustained activity that is also genuinely needed for creativity. As indicated above, the particular degree of emphasis on the timeless and the temporal orders will have to vary with the particular context. A proper attention to this question clearly opens up a new order of possibilities for creativity.

There is a commonly accepted view that it is not actually possible to say anything relevant about creativity, because to do this would necessarily limit the latter in a mechanical and therefore uncreative way. However, this viewpoint is extreme and has the effect of preventing any inquiry into the question of what impedes creativity and how creativity could be fostered. Indeed the rigid extremes of the timeless and the temporal, or the verbal and the nonverbal, are precisely one of the major blocks to creativity itself. Therefore it is possible that something relevant may be said about creativity, provided it is realized that whatever we say it is, there is also something more and something different. In this way any discussion of creativity acts as a point of departure rather than a definitive statement about "what is."

This section has in fact gone beyond the normal scope of discourse in the book so far, which has mainly dealt in terms of facts and ideas. It has investigated and suggested approaches that may be favorable to creativity, or at least to "loosening" the rigid, tacit infrastructures that impede its operation. In the next chapter this approach will be developed even further to indicate a number of key approaches that may favor the flowering of creativity.

SUMMARY AND CONCLUSIONS

This chapter brought together science, nature, society, and consciousness in terms of an overall common generative order. Creativity was found to act, not only through the free play of thought but also through a similar free movement of awareness and attention. These are what make possible the action of creative intelligence which unfolds from its indefinable subtle source toward manifestation down the "stream" of the generative order.

The question was then raised as to whether computers will ever adequately simulate this action. It was suggested that they should instead be regarded as providing a kind of *artificial* intellect rather than *artificial intelligence*. Finally the question of how the timeless generative order and the temporal order of succession may be related was explored and it was seen that an immense new creative domain of order lies between these two extremes.

SIX

CREATIVITY IN THE WHOLE OF LIFE

In the previous chapter it was suggested that the potential for creativity is natural but that an excessively rigid attachment to fixed "programs" in the tacit infrastructure of consciousness is primarily what prevents this creativity from acting. The present chapter will explore the nature of these blocks in greater detail, and the social and cultural conditioning that lies behind them. This leads to a discussion of free dialogue, which is proposed as a key activity in which this sort of conditioning can be dissolved. Such free dialogue is fundamentally relevant to the whole question of how conditioning of the individual and of humanity, to falseness at the sociocultural level, can come to an end.

This chapter also contrasts the approaches of East and West to these important questions and suggests that a new order is in fact needed which goes beyond both. Finally the overall planetary culture is discussed in terms of three basic attitudes of mind to the whole of life: the scientific, the artistic, and the religious. The main purpose of the book is then summed up in a discussion suggesting what would be involved in a general release of human creative energies.

CREATIVITY AND WHAT BLOCKS IT

In the Introduction, a call was made for a new surge of creativity in science. By now it will be clear that such a surge must extend into all areas of human activity if the actual challenge, which has finally revealed itself, is to be met. But does this mean that creativity must somehow be elicited from an organism that does not have in itself a natural potential for creativity? It is proposed that, on the contrary, human beings do indeed have such a potential. However, as children grow older, this creativity appears to be blocked.

Some insight into the nature of this block can be gained from the work of Desmond Morris, published in *The Biology of Art*.[1] In one experiment chimpanzees were given canvas and paint and immediately began to apply themselves to make balanced patterns of color, somewhat reminiscent of certain forms of modern art, such as abstract expressionism. The significant point about this experiment is that the animals became so interested in painting and it absorbed them so completely that they had comparatively little interest left for food, sex, or the other activities that normally hold them strongly. Additional experiments showed somewhat similar results for other primates. When very young children are given paints, their behavior is remarkably like that of the chimpanzees.

This seems to indicate that creativity is a natural potential. Yet somehow, in most cases, the urge to create fades as the human being gets older. Or at best it continues in certain limited areas, such as science, music, or painting. Why should this happen?

An extension of Morris's experiment involved rewarding the chimpanzees for producing their paintings. Very soon their work began to degenerate until they produced the bare minimum that would satisfy the experimenter. A similar behavior can be observed in young children as they become "self-conscious" of the kind of painting they believe they are "supposed" to do. This is generally indicated to them by subtle and implicit rewards, such as praise and approval, and by the need

1. Methuen, London, 1962.

to conform to what other children around them are doing. Thus creativity appears to be incompatible with external and internal rewards or punishments. The reason is clear. In order to do something for a reward, the whole order of the activity, and the energy required for it, are determined by arbitrary requirements that are extraneous to the creative activity itself. This activity then turns into something mechanical and repetitious, or else it mechanically seeks change for its own sake. The state of intense passion and vibrant tension that goes with creative perception in the way discussed in Chapter 1 then dies away. The whole thing becomes boring and uninteresting, so that the kind of energy needed for creative perception and action is lacking. As a result, even greater rewards, or punishments, are needed to keep the activity going.

Basically, the setting of goals and patterns of behavior, which are imposed mechanically or externally, and without understanding, produces a rigid structure in consciousness that blocks the free play of thought and the free movement of awareness and attention that are necessary for creativity to act. But this does not mean that rules and external orders are incompatible with creativity, or that a truly creative person must live in an arbitrary fashion. To write a sonnet or a fugue, to compose an abstract painting, or to discover some new theorem in mathematics requires that creativity should operate within the context of a particular artistic or mathematical form. Cézanne's particular creativity in art, for example, was directed toward the discovery of new forms and orders of composition within the context of a particular form of freedom that had been previously established by the Impressionists. Some of Bach's greatest works are similarly created within the confines of strict counterpoint. To live in a creative way requires extreme and sensitive perception of the orders and structures of relationship to individuals, society, and nature. In such cases, creativity may flower. It is only when creativity is made subservient to external goals, which are implied by the seeking of rewards, that the whole activity begins to wither and degenerate.

Whenever this creativity is impeded, the ultimate result is not simply the absence of creativity, but an actual positive presence of destructiveness, as was suggested in Chapter 5. In the case of the painting experiment, this shows up as a false attitude. Both the chimpanzee and the child are engaged in an activity that no longer has meaning in itself, merely in order to experience a pleasant and satisfying state of consciousness, in the form of reward or the avoidance of punishment. This introduces something that is fundamentally false in the generative order of consciousness itself. For example, the continuation of this approach would eventually lead the child to seek pleasing words of praise from others, even if they are not true, and to collude with others in exchanging flattering remarks that lead to mutual satisfaction. This, however, is achieved at the expense of self-deception that can, in the long run, be quite dangerous.

What is even of greater danger to the child, in such an approach, is that it eventually brings about violence of various kinds. For creativity is a prime need of a human being and its denial brings about a pervasive state of dissatisfaction and boredom. This leads to intense frustration that is conducive to a search for exciting "outlets," which can readily involve a degree of force that is destructive. This sort of frustration is indeed a major cause of violence. However, what is even more destructive than such overt violence is that the senses, intellect, and emotions of the child gradually become deadened and the child loses the capacity for free movement of awareness, attention, and thought. In effect, the destructive energy that has been aroused in the mind has been turned against the whole creative potential itself.

Most education does in fact make use, in explicit or in more hidden and subtle ways, of rewards and punishments as key motivating factors. For example, the whole philosophy of behavior modification and positive reinforcement, which is particularly prevalent in North American education, holds that a system of rewards is essential for effective learning. This alone is a tremendous barrier to creativity.

In addition, education has traditionally given great value to fixed knowledge and techniques. In this way it places an extremely great importance on authority as determining the very generative order of the psyche. What is involved is not only the authority of the teacher as a source of knowledge that is never to be questioned, but even more, the general authority of knowledge itself, as a source of truth that should never be doubted. This leads to a fundamental loss of self-confidence, to a blockage of free movement and a corresponding dissipation of energy, deep in the generative order of the whole of consciousness. Later on, all of this may show up as a disposition to be afraid of inquiring into fundamental questions, and to look to experts and "geniuses" whenever any difficulty or basic problem is encountered.

Of course, a certain reasonable kind of authority is needed to maintain necessary order in the classroom. And the student has to realize that, in broad areas, the teacher has valuable knowledge that can be conveyed in an appropriate way. But what is important is the overall *attitude* to this knowledge. Does it seek to impose itself arbitrarily and mechanically deep within the generative order of the mind, or does it allow itself to be discussed and questioned, with a view to making understanding possible? Similar questions can be raised with regard to conformity to arbitrary norms, which come not only from the teacher, but even more from the peer group and from society at large.

Beyond school, society operates in much the same way, for it is based largely on routine work that is motivated by various kinds of fear and by arbitrary pressures to conform as well as by the hope for rewards. Moreover, society generally regards this as necessary and valuable and, in turn, treats creativity as irrelevant for the most part, except in those special cases, such as science and the arts, in which it is rewarded. In fact, no society has thus far managed to organize itself in a complex way without using a system of rewards and punishments as a major inducement to bring about cooperation. It is generally felt that if society tried to do without these, whether in the family, in the classroom, at work, or in broader contexts, it would incur the risk of eventual total disruption and chaos. Creativity is nevertheless a major need of each human being and the blockage of this creativity eventually threatens civilization with ultimate destruction.

Humanity is therefore faced with an urgent challenge of unparalleled magnitude. Specifically, rigidity in the generative order, to which control through rewards and punishments makes a major contribution, prevents the free play of thought and the free movement of awareness and attention. This leads to false play which ultimately brings about a pervasive destructiveness while at the same time blocking natural creativity of human beings.

A proper response to this challenge requires the kind of overall creativity in society that is implicit in the call being made in this book for a general creative surge in all areas of life. Clearly from this it would follow that the various forms of rigidity that have already been discussed would all change fundamentally. But such a change cannot be restricted to a single overall flash of insight. Creativity has to be *sustained*. For example, in Chapter 4 it was shown how the artist has to work constantly from the creative source in the generative order. An artist does not have a creative vision and *then* apply it mechanically, in a sequential process by means of rules, techniques, and formulae. Rather, these latter flow out of the sustained creative vision in a creative way.

To pay serious attention to this need for sustained creativity is extremely relevant for bringing about a creative change in culture and in society. In most cases, however, creative new discoveries are generally followed by an attempt to reduce them to something that can be applied mechanically. While mechanical application is necessary in certain contexts, the basic impetus for each individual must come from the creative origin, and this is beyond any mechanical, explicate, or sequential order of succession.

It is possible to point to specific areas in which a creative change would be of great benefit to society and the individual. For example, by means of a tremendous creative common action, education must no longer depend on rewards and punishments, no matter how subtle these may be. It must also cease to place an excessively high value on arbitrary authority, fixed knowledge, and techniques and conformity. Some partial and preliminary work in this direction has been done from time to time. For example, there has been an effort to present the child with a great deal of meaningful material to arouse interest, so that the child does not have to be offered a reward to learn. Also, some people working in this field have emphasized free play as a way of arousing creativity. Others have given much attention to relationships that avoid unnecessary authority and conformity. By the further development of such approaches, it should in principle be possible for children to learn without the inducement of rewards.

However, there are deeper difficulties, which prevent these approaches from actually working in the long run. The problem does not stem primarily from the field of education alone. Rather, it arises ultimately out of the tacit infrastructure of the entire consciousness of humanity. This is deeply and pervasively conditioned, for example, by general tradition that takes the absolute necessity of rewards and punishments for granted. Both teachers and students are caught up in subtler forms of the same false structure that they are explicitly trying to avoid. This may, in the long run, be at least as destructive as was the original pattern that the whole experiment in education was designed to avoid.

It seems that the whole conditioning of all who take part must in fact change: society, the family, and the individual. It is thus clear that there is no *single* stationary point at which these problems might be attacked. The educational system, society, and the individual are all intimately involved. But it is ultimately the overall order of human consciousness that has to be addressed.

BLOCKS TO CREATIVITY IN THE GENERATIVE ORDER OF SOCIETY

Creativity, in almost every area of life, is blocked by a wide range of rigidly held assumptions that are taken for granted by society as a whole. Some of these have already been discussed in this book, but in addition, every society holds additional assumptions that are of such a shaky nature that they are not even admitted into discussion. There is therefore an unspoken requirement that everyone must subscribe to these assumptions, but that no one should ever mention that any such assumptions indeed exist. They are tacitly denied as operating within society, and even this denial is denied. The overall effect is to lead people to collude in "playing false" so they constantly distort all sorts of additional thoughts in order to protect these assumptions. Such bad faith enters deep into the overall generative order of society.

These rigidities and fixed assumptions, many of which must not be mentioned but must nevertheless be defended, may be compared with a kind of pollution that is constantly being poured into the stream of the generative order of society. It makes no sense to attempt to "clean up" parts of this pollution farther downstream while continuing to pollute the source itself. What is needed is either to stop the pollution at its source, or to introduce some factor into the stream that naturally "cleans up" pollution.

In the body a similar problem arises. As a person grows older, through infection, allergies, contaminants, misadventure, and the processes of aging, considerable "misinformation" or irrelevant information accumulates in the system. Indeed it is possible to look at a disease like cancer as arising from misinformation in the structure of DNA. Viruses also introduce misinformation, in the sense that DNA from the virus acts to replace some of the DNA in the host cell and therefore causes this cell to replicate foreign DNA rather than serving the needs of the body.

There are basically three ways of dealing with this problem of misinformation in the body. The first is to avoid the introduction of misinformation in the first place, for example, by keeping away from infection through good sanitation and a careful diet. Second, where misinformation exists, it may be possible to do something to remove it through various kinds of medical intervention. But more significantly, the third option involves the body itself, which possesses an immune system which is able to "clear up" misinformation in a natural way.

This is indeed the body's main mode of dealing with misinformation. This can be clearly seen from the fact that drugs are of little use in treating a disease like AIDS, which destroys this immune system itself. Furthermore, the whole practice of immunization relies on activating the immune system and so avoiding the onset of particular diseases.

The immune system itself is particularly complex and contains a very subtle kind of information that can respond to the whole "meaning" of what is happening to the order of the body. In this way it is able to distinguish misinformation from information needed for the body's healthy operation. It can be compared to a kind of "intelligence" that works within the body. Moreover there is evidence that this sort of "intelligence" can respond to the higher levels that are usually associated with thought and feeling. It is well known that depressing thoughts can inhibit the activity of the immune system, with the result that a person becomes more susceptible to infections. Indeed there is much evidence that a vigorous, creative state of mind and a strong "will to live" are conducive to general health and even to recovery from dangerous illnesses.[2] More generally, it could be said that good health is basically a manifestation of the overall creative intelligence, working in concert with the body, through various means that include exercise, diet, relaxation, and so on.

Returning to a consideration of society, clearly there is also a vast amount of misinformation in circulation which acts toward society's degeneration. The media and various modern means of communication have the effect of rapidly disseminating and magnifying this misinformation, just as they do with valid information. It should be clear that by "misinformation" is meant a form of *generative information* that is inappropriate, rather than simply incorrect statements of fact. In a similar way a small "mistake" in DNA can have disastrous consequences because it forms part of the generative order of the organism and may set the whole process in the wrong direction.

2. See, for example, Norman Cousins, *Anatomy of an Illness*, Bantam, New York, 1981.

In society, the generative order is deeply affected by what has a very *general* significance. Indeed the generative order may be regarded as the *concrete activity of the general*. This takes the form of general principles, general aims, and generally accepted values, attitudes, and beliefs of all kinds that are associated with the family, work, religion, and country. In going from these general principles to the universal, it is clear that the effect on the generative order will become yet more powerful. When a given principle is regarded as universally valid, it means that it is taken as absolutely necessary. In other words, things cannot be otherwise, under any circumstances whatsoever. Absolute necessity means "never to yield." To have something in the generative order that can never give way, no matter what happens, is to put an absolute restriction on free play of the mind, and thus to introduce a corresponding block to creativity that is very difficult to move.

Of course, both the individual and society require a certain stability, and for this, thought must be able to hold itself fixed within certain appropriate limits and with a certain kind of *relative* necessity.

Over a limited period of time, certain values, assumptions, and principles may usefully be regarded as necessary. They are relatively constant, although they should always be open to change when evidence for the necessity of the latter is perceived. The major problem arises, however, when it is assumed, usually tacitly and without awareness and attention, that these values, assumptions, and principles have to be absolutely fixed, because they are taken as necessary for the survival and health of the society and for all that its members hold to be dear.

In the beginning of this book it was argued that science, which is in principle dedicated to the truth, tends to be caught up in necessity which then leads to false play and a serious blockage of creativity. It is now clear that the assumptions of absolute necessity, with their predispositions to unyielding rigidity, are only part of a much broader spectrum of similar responses that pervade society as a whole. General principles, values, and assumptions, which are taken in this way to have absolute necessity, are thus seen as a major source of the destructive misinformation that is polluting the generative order of society.

As with the body, society attempts to deal with this sort of misinformation by trying to prevent it from entering its fabric, or attempting to "cure" it with some form of therapy. For example, on a rather superficial level, there are laws to prevent false information and information which may engender hatred, anger, and prejudice from being spread about various races, religions, and groups. Writers, dramatists, and filmmakers go some way to making people aware of prejudices and rigidly fixed attitudes. But in the long run, all these attempts are limited by the overwhelming, and yet often very subtle, pressures within society toward colluding to defend one's own group and its ideas. In addition, there is the whole problem of the intolerance and mistrust that have grown up between nations, religions, ideologies, and other groups which go all the way down to the family itself. To some extent psychotherapy and group therapies can help to clear up individual misinformation of this kind, which may go back to early childhood, or start in a later phase of life. But these approaches have very little effect in the larger sphere of society as a whole.

A particularly important piece of misinformation is the key assumption that creativity is necessary only in specialized fields. This assumption pervades the whole culture, but most people are generally not aware of it; there is always a tendency for misinformation to defend itself by leading people to collude in playing false, whenever such an assumption is questioned. Assuming the restricted nature of creativity is obviously of serious consequence for it clearly predetermines any program that is designed to clear up the misinformation within society and suggests that it cannot be creative.

All that seems to be left is to ask whether society contains some kind of "immune system" that could spontaneously and naturally clear up misinformation. If such a system exists, then it is certainly not obvious, nor does it appear to be in common operation within our society today.

DIALOGUE AND CULTURE

In this section it is proposed that a form of free dialogue may well be one of the most effective ways of investigating the crisis which faces society, and indeed the whole of human nature and consciousness today. Moreover, it may turn out that such a form of free exchange of ideas and information is of fundamental relevance for transforming culture and freeing it of destructive misinformation, so that creativity can be liberated. However, it must be stressed that what follows is not given in the spirit of a prescription that society is supposed to follow. Rather it is an invitation to the reader to begin to investigate and explore in the spirit of free play of ideas and without the restriction of the absolute necessity of any final goal or aim. For once necessity and absolute requirements or directions enter into the spirit of this exploration, then creativity is limited and all the problems that have plagued human civilization will surface yet again to overwhelm the investigation.

To begin, it should be noted that many of the ideas to be explored were first investigated by Patrick de Maré, who is a psychiatrist working in England.[3] De Maré has used his wide experience of dialogue in therapeutic groups to support his arguments. However, it is essential to emphasize that his ideas about dialogue are not concerned primarily with psychotherapy, but rather with the transformation of culture, along the general lines that have been indicated in this chapter.

In the first two chapters it was shown how rigid conditioning of the tacit infrastructure of scientific thought has led to a fragmentation in science and to an essential breakdown in communication between areas which are considered to be mutually irrelevant. Nevertheless a closer investigation of actual cases suggested that there is nothing inherent in science which makes such breaks in communication and fragmentation inevitable. Indeed wherever fragmentation and failures in communication arise, this clearly indicates that a kind of dialogue should be established.

3. A brief, but fairly comprehensive presentation of de Maré's ideas can be found in *Group Analysis*, vol. XVII, no. 78, Sage, London, 1985.

The term *dialogue* is derived from a Greek word, with *dia* meaning "through" and *logos* signifying "the word." Here "the word" does not refer to mere sounds but to their meaning. So dialogue can be considered as a free flow of meaning between people in communication, in the sense of a stream that flows between banks.

A key difference between a dialogue and an ordinary discussion is that, within the latter, people usually hold relatively fixed positions and argue in favor of their views as they try to convince others to change. At best this may produce agreement or compromise, but it does not give rise to anything creative. Moreover, whenever anything of fundamental significance is involved, then positions tend to be rigidly nonnegotiable and talk degenerates either into a confrontation in which there is no solution, or into a polite avoidance of the issues. Both these outcomes are extremely harmful, for they prevent the free play of thought in communication and therefore impede creativity.

In dialogue, however, a person may prefer a certain position but does not hold to it nonnegotiably. He or she is ready to listen to others with sufficient sympathy and interest to understand the meaning of the other's position properly and is also ready to change his or her own point of view if there is good reason to do so. Clearly a spirit of goodwill or friendship is necessary for this to take place. It is not compatible with a spirit that is competitive, contentious, or aggressive. In the case of Einstein and Bohr, which was discussed in Chapter 2, these requirements were evidently met, at least initially. However, because each felt that a different notion of truth and reality was involved, which was not negotiable in any way at all, a real dialogue could never take place.

This brings us to an important root feature of science, which is also present in dialogue: to be ready to acknowledge any fact and any point of view as it actually is, whether one likes it or not. In many areas of life, people are, on the contrary, disposed to collude in order to avoid acknowledging facts and points of view that they find unpleasant or unduly disturbing. Science is, however, at least in principle, dedicated to seeing any fact as it is, and to being open to free communication with regard not only to the fact itself, but also to the point of view from which it is interpreted. Nevertheless, in practice, this is not often achieved. What happens in many cases is that there is a blockage of communication.

For example, a person does not acknowledge the point of view of the other as being a reasonable one to hold, although perhaps not correct. Generally this failure arises when the other's point of view poses a serious threat to all that a person holds dear and precious in life as a whole.

In dialogue it is necessary that people be able to face their disagreements without confrontation and be willing to explore points of view to which they do not personally subscribe. If they are able to engage in such a dialogue without evasion or anger, they will find that no fixed position is so important that it is worth holding at the expense of destroying the dialogue itself. This tends to give rise to a unity in plurality of the kind discussed in Chapter 3. This is, of course, quite different from introducing a large number of compartmentalized positions that never dialogue with each other. Rather, a plurality of points of view corresponds to the earlier suggestion that science and society should consist not of monolithic structures but rather of a dynamic unity within plurality.

One of the major barriers to this sort of dialogue is the rigidity in the tacit infrastructure of the individual and society, which has been discussed throughout this book. The tacit infrastructure of society at large is contained in what is generally called culture. Within each society, however, there are many subcultures which are all somewhat different, and which are either in conflict with each other, or more or less ignore each other as having mutually irrelevant aims and values. Such subcultures, along with the overall culture, are generally rigidly restricted by their basic assumptions, most of which are tacit and not open to awareness and attention. Creativity is therefore, at best, an occasional occurrence, the results of which are quickly absorbed in a fairly mechanical way into the general tacit infrastructure.

At present, a truly creative dialogue, in the sense that has been indicated here, is not at all common, even in science.

Rather the struggle of each idea to dominate is commonly emphasized in most activities in society. In this struggle, the success of a person's point of view may have important consequences for status, prestige, social position, and monetary reward. In such a conditioned exchange, the tacit infrastructure, both individually and culturally, responds very actively to block the free play that is needed for creativity.

The importance of the principle of dialogue should now be clear. It implies a very deep change in how the mind works. What is essential is that each participant is, as it were, suspending his or her point of view, while also holding other points of view in a suspended form and giving full attention to what they mean. In doing this, each participant has also to suspend the corresponding activity, not only of his or her own tacit infrastructure of ideas, but also of those of the others who are participating in the dialogue. Such a thoroughgoing suspension of tacit individual and cultural infrastructures, in the context of full attention to their contents, frees the mind to move in quite new ways. The tendency toward false play that is characteristic of the rigid infrastructures begins to die away. The mind is then able to respond to creative new perceptions going beyond the particular points of view that have been suspended.

In this way, something can happen in the dialogue that is analogous to the dissolution of barriers in the "stream" of the generative order that was discussed at the end of the previous chapters. In the dialogue, these blockages, in the form of rigid but largely tacit cultural assumptions, can be brought out and examined by all who take part. Because each person will generally have a different individual background, and will perhaps come from a different subculture, assumptions that are part of a given participant's "unconscious" infrastructure may be quite obvious to another participant, who has no resistance to seeing them. In this way the participants can turn their attention more generally to becoming aware, as broadly as possible, of the overall tacit infrastructure of rigid cultural and subcultural assumptions and bringing it to light. As a result, it becomes possible for the dialogue to begin to play a part that is analogous to that played by the immune system of the body, in "recognizing" destructive misinformation and in clearing it up. This clearly constitutes a very important change in how the mind works.

There is, however, another extremely important way in which the operation of the mind can be transformed in such a dialogue. For when the rigid, tacit infrastructure is loosened, the mind begins to move in a *new order*. To see the nature of this order, consider first the order that has traditionally characterized cultures. Essentially this involves a strong fragmentation between individual consciousness—"what the individual knows all together"—and social consciousness—"what the society knows all together."

For the individual, consciousness tends to emphasize subjectivity in the sense of private aims, dreams, and aspirations that are shared to some extent with family and close friends, as well as a general search for personal pleasure and security. In society, however, consciousness tends to emphasize a kind of objectivity with common aims and goals, and there is an attempt to put conformity and the pursuit of the common welfare in the first place. One of the principal conflicts in life arises therefore in the attempt to bring these two fragments together harmoniously. For example, as a person grows up, he (or she) may find that his individual needs have little or no place in society. And in turn, as society begins to act on the individual consciousness in false and destructive ways, people become cynical. They begin to ignore the requirements of reality and the general good in favor of their own interests and those of their group.

Within this generally fragmentary order of consciousness, the social order of language is largely for the sake of communicating information. This is aimed, ultimately, at producing results that are envisaged as necessary, either to society or to the individual, or perhaps to both. Meaning plays a secondary part in such usage, in the sense, for example, that what are put first are the problems that are to be solved, while meaning is arranged so as to facilitate the solution of these problems. Of course, a society may try to find a common primary meaning in myths, such as that of the invincibility of the nation or its glorious destiny. But these lead to illusions, which are in the long run unsatisfactory, as well as dangerous and destructive. The individual is thus generally left with a desperate search for something that would give life real meaning. But this can seldom be found either in the rather crude mechanical, uncaring society, or in the isolated and consequently lonely life of the individual. For if there is not common meaning to be shared, a person can be lonely even in a crowd.

What is especially relevant to this whole conflict is a proper understanding of the nature of culture. It seems clear that in essence culture *is* meaning, as shared in society. And here "meaning" is not only *significance* but also *intention, purpose*, and *value*. It is clear, for example, that art, literature, science, and other such activities of a culture are all parts of the common heritage of *shared meaning*, in the sense described above. Such cultural meaning is evidently not *primarily* aimed at utility. Indeed, any society that restricts its knowledge merely to information that it regards as useful would hardly be said to have a culture, and within it, life would have very little meaning. Even in our present society, culture, when considered in this way, appears to have a rather small significance in comparison to other issues that are taken to be of vital importance by many sectors of the population.

The gulf between individual consciousness and social consciousness is similar to a number of other gulfs that have already been described in this book, for example, between descriptive and constitutive orders, between simple regular orders of low degree and chaotic orders of infinite degree, and, of course, between the timeless and time orders. But in all these cases, broad and rich new areas for creativity can be found by going to new orders that lie *between* such extremes. In the present case, therefore, what is needed is to find a broad domain of creative orders between the social and individual extremes. Dialogue therefore appears to be a key to the exploration of these new orders.

To see what is involved, note that as the above dialogue develops, not only do specific social and cultural assumptions "loosen up," but also much deeper and more general assumptions begin to be affected in a similar way. Among these, one of the most important is the assumption that between the individual consciousness and the social consciousness there is an absolute gulf. This implies that the individual must adjust to fit into the society, that society must be remade to suit the individual, or that some combination of both approaches must be carried out. If, however, the dialogue is sustained sufficiently, then all who participate will sooner or later be able to see, in actual fact, how a creative movement can take place in a new order between these extremes. This movement is present both externally and publicly, as well as inwardly, where it can be felt by all. As with alert attention to a flowing stream, the mind can then go into an analogous order. In this order, attention is no longer restricted to the two extreme forms of individual and social. Rather, attention is transformed so that it, along with the whole generative order of the mind, is in the rich creative domain "between" these two extremes.

The mind is then capable of new degrees of subtlety, moving from emphasis on the whole group of participants to emphasis on individuals, as the occasion demands. This is particularly significant for proper response to the strong emotional reactions that will inevitably arise, even in the friendliest group, whenever fundamental assumptions are disturbed. Because the mind is no longer rigidly committed to the individual or to the social extremes, the basic issues that arise in a disagreement between participants are to a considerable extent "defused." For the assumptions that are brought to the common attention are no longer implied to have absolute necessity. And as a result, the "emotional charge" that is inevitably associated with an assumption that is dear to one or more members of the group can be reduced to more manageable proportions, so that violent "explosions" are not likely to take place. Only a dialogue that can, at the same time, meet the challenge both of uncovering the intellectual content of a rigidly held basic assumption and of "defusing" the emotional charge that goes with it will make possible the proper exploration of the new order of mental operation that is being discussed here.

It is possible to have such dialogues in all sorts of circumstances, with many or just a few people involved. Indeed even an individual may have a kind of internal dialogue with himself or herself. What is essential here is the presence of the *spirit* of dialogue, which is, in short, the ability to hold many points of view in suspension, along with a primary interest in the creation of a common meaning. It is particularly important, however, to explore the possibilities of dialogue in the context of a group that is large enough to have within it a wide range of points of view, and to sustain a strong flow of meaning. This latter can come about because such a dialogue is capable of having the powerful nonverbal effect of consensus. In the ordinary situation, consensus can lead to collusion and to playing false, but in a true dialogue there is the possibility that a new form of consensual mind, which involves a rich creative order between the individual and the social, may be a more powerful instrument than is the individual mind. Such consensus does not involve the pressure of authority or conformity, for it arises out of a spirit of friendship dedicated to clarity and the ultimate perception of what is true. In this way the tacit infrastructure of society and that of its subcultures are not opposed, nor is there any attempt to alter them or to destroy them. Rather, fixed and rigid frames dissolve in the creative free flow of dialogue as a new kind of microculture emerges.

People who have taken part in such a dialogue will be able to carry its spirit beyond the particular group into all their activities and relationships and ultimately into the general society. In this way, they can begin to explore the possibility of extending the transformation of the mind that has been discussed earlier to a broader sociocultural context. Such an exploration would clearly be relevant for helping to bring about a creative and harmonious order in the world. It should be clear by now that the major barriers to such an order are not technical; rather they lie in the rigid and fragmentary nature of our basic assumptions. These keep us from changing in response to the actual situations and from being able to move together from commonly shared meanings.

THE INDIVIDUAL, THE SOCIAL, AND THE COSMIC DIMENSION OF THE HUMAN BEING

Dialogue, in the sense that has been described here, may be able to contribute in a very significant way to clearing up the "pollution" or "misinformation" in social and cultural spheres. But humanity does not live *only* in these spheres. Broadly speaking it has three principal kinds of dimension—the individual, the social, and the cosmic—and each of these must receive its appropriate attention.

Consider first how conditioning operates in the individual dimension. Each individual, throughout his or her life, accumulates from society at large a vast amount of misinformation in the generative order. This individual misinformation is perhaps described by the word *idiosyncrasy*, whose Greek root means "private mixture." Each human being has thus to address his or her own particular "mixture," which has been built up since the day of birth.

One of the most important ways of dealing with individual problems, which has developed in relatively recent times, involves psychiatry and allied approaches. In this field, the work of Sigmund Freud stands out as playing a seminal role. It is based on the observation that neuroses in adults have their roots in experiences of childhood and infancy that were never understood properly, and were so painful that they were repressed from conscious awareness. Such repressed material evidently corresponds in some sense to the notion of a rigid tacit infrastructure, which is largely unconscious and tends to lead to false play and the blockage of creativity. Freud's treatment essentially was to try to bring this repressed material into conscious awareness so that the misinformation in the generative order could be corrected or spontaneously dropped.

Freud's approach, and that of a large number of other psychiatrists who followed him, was to put a major emphasis on clearing up misinformation acquired in early childhood. Clearly, common sense would imply that what happens in the first few years of life must be very important in this regard. As the saying goes, "As the twig is bent, so grows the tree."

However, a number of psychiatrists have felt that Freud concentrated too exclusively on the early years and paid too little attention not only to what comes later but also to other factors, such as society, the culture, and the general inheritance of the human race as a whole, for example, Jung's archetypes and collective unconscious. Nevertheless, in spite of such disagreements, it is clear that the psychoanalytic approach has achieved worthwhile results and insights.

However, in the context of this book, it is necessary to question Freud's notion that creativity, especially in the intellectual spheres, arises essentially from a displacement or sublimation of what he calls the *libido*. According to the dictionary, this latter concept means pleasure, desire, eager longing, and sensual passion. Freud for his part gave it a more extended meaning so that it implies a kind of general mental energy, at first directed toward sex and later invested in any object that is significant to the ego.[4] Freud's theory argues that when this mental energy is seriously frustrated early in life, it is turned or displaced toward some other outlet, such as art or science. Of course, there can hardly be any doubt that something like this actually happened to a number of leading scientists and artists. For example, Newton suffered a difficult childhood and it is plausible that this led him to turn to science by way of compensation.

However, this is clearly not an explanation of what creativity actually *is*, and how it originates. It could hardly be said that the libido, in itself, is creative. Indeed, insofar as it may lead to an excessively strong investment of mental energy in certain fixed objects, it will tend to bring about the kind of rigidity that interferes with creativity.

Rather, when early in life, the natural mental energy finds it impossible to respond creatively in relationships within the family, it may turn toward other areas in which this can be done. In a limited sense, this is a kind of "solution," though clearly a much more thoroughgoing and pervasive kind of

4. See Charles Rycroft, *A Critical Dictionary of Psychoanalysis*, Nelson, London, 1968.

creative response is needed in the long run. However, what may happen with such individuals who are called geniuses is that they manage somehow to sustain creativity throughout at least a significant part of their lives, albeit in some limited area, whereas with most people this is largely "damped down" by various social mechanisms, such as rewards and punishments. It is the view of this book that the potential for creative perception is natural. Creativity originates in the depths of the generative order, and the proper role of mental energy is to respond to such perception, and ultimately to bring it to some manifest level of reality. One of the main functions of psychiatry should be to free creativity from rigid blocks within the individual, whether these begin in early childhood or in other contexts.

Such a proposal, however, appears to run counter to what is commonly taken as the main aim of psychiatry, namely, to help the individual to adjust to society. This would perhaps have seemed to have made some kind of sense in Freud's time, when people believed society to be basically sound and healthy. Following two world wars, however, the decay of society has become so evident that many psychiatrists are no longer satisfied with this sort of aim. One of the most outspoken is R. D. Laing, who feels that what is called insanity can actually be a "sane" response to a "mad" society. De Maré for his part does not take such an extreme position but calls for a creative transformation of culture through dialogue, which will deeply affect both the individual and the society together.

De Maré has also pointed out a close similarity between a free dialogue, with no fixed tasks or goals, and Freud's method of free association in psychoanalysis, which helps to bring repressed content into awareness. In our approach, both of these can be considered as examples of how the mind can begin to move in new ways that are not bound by its rigid conditioning. With psychoanalysis, it is mainly the individual's particular conditioning that is revealed. But in dialogue, what is revealed is primarily the cultural conditioning.

Both kinds of conditioning are mainly "unconscious," for since the mind defends itself by various forms of self-deceptive false play, it is not able to give awareness and attention to the nature of its own conditioning. In addition, just as a person is not aware of his or her accent, so does cultural conditioning escape awareness in this subtler sense. The general cultural conditioning is probably in the long run even more powerful than that which originates in the early years of life. Moreover, the usual psychiatric approach has little or no impact on basic cultural assumptions, which are as likely to be "unconsciously" held by the psychiatrist as by the patient.

In a free dialogue, however, with many different individuals representing a variety of subcultures, all having a specific interest in becoming aware of rigid cultural assumptions, a new order of operation of the mind between the individual and society can develop. In this way cultural conditioning can be dissolved in a dialogue in which the participants operate between the individual and social dimensions of the human being.

Having discussed the individual and social dimensions of the human being, it is now time to turn to the cosmic dimension. This is concerned with the human relationship to the whole, to the totality of *what is*. From the earliest times it has been considered crucial, for the overall order of the individual and society, that a harmonious relationship be established with this whole. Indeed it was commonly believed that such a relationship would serve to prevent or dissolve the various sorts of difficulties that have been discussed in this book in connection with destructive "misinformation" and with the tacit infrastructure of consciousness.

In the very distant past, human beings obtained their sense of harmony within the cosmic dimension through direct contact with nature. When people were constantly immersed in their natural environment, their attention naturally turned in this direction and consciousness frequently moved into a dimension beyond time and the limited concerns of particular social groups. Even now, when people spend some time close to nature they may experience something of this "healing" quality in body and mind. In earlier times humans were in almost constant contact with nature so that "misinformation" arising, for example, from social contacts would have little or no ultimate significance, as it was constantly "washed away."

However, as civilization developed, this immediate contact with nature grew more tenuous. To some extent it was replaced by philosophy and science, which also gave human beings a certain sense of relationship to the totality. But as science developed into ever more abstract and institutionalized structures, the sense of contact became more and more indirect and restricted to limited groups of specialists who understood the highly mathematical theories. While specialists had the skill to use the complex instruments of theory and experiment to mediate between nature and human beings, for the vast majority of people such contact was superficial and indirect. In general it is now restricted to the writings of those who try to translate the mathematical abstractions of physics into a nontechnical language.

Another approach to the totality, which was present even in the very earliest times, was through religion. As civilization developed and grew more separate from nature, religion probably became more and more important. By the time of the Middle Ages, for example, religion was the main means by which Europeans maintained a sense of contact with the whole. But with the coming of the modern era, science began to make this religious worldview appear implausible to many people. Today in both East and West, religion has, by and large, ceased to be the principal source of the ultimate meaning in life. Yet science, for its part, has been unable to take its place in this regard.

Western religions emphasize belief in the Supreme Being as the source not only of the cosmos but of all that makes life worthwhile for human beings in the cosmos that this Being, or God, has created. In Eastern religions, gods have played an important role, but the general development has been toward discovering the ultimate ground of all being. In the Indian religions, for example, the fundamental source has been named Brahman, and the key insight is that the ultimate self, or Atman, is identical with the ultimate being, or Brahman.

As a verbal statement, however, this means very little by

itself. The essential point of such a religious and philosophical attitude is to enter directly into the absolute reality (or Brahman). This frees the individual of all "misinformation" in the overall conditioning of the tacit infrastructure of consciousness and brings about a state of ultimate bliss and perfection. Only a few have ever claimed to have achieved this, but vast numbers of people have been profoundly affected by such notions. In the West, mystics have a somewhat similar notion of union with God or the Godhead, but the emphasis has tended to be placed more on grace rather than on individual perception and understanding.

For most people, however, religion is not primarily a question of mysticism. Rather, for them the emphasis lies in some form of belief in a Supreme Being and in a set of principles and practices that follow from such a belief, and are used in daily living. (However, it must be added that in Buddhism, there is no belief in a Supreme Being; rather the emphasis is placed on proper understanding of the ultimate groundlessness of the "self.")

The principal difficulty with the religious approach, and indeed with any attempt to make a formal definition of the totality and of an individual's relationship with it, is that it tends very strongly to produce rigidly fixed ideas. These are very heavily emotionally charged so that they prevent the free play of the mind, and thus bring about destructive false play and the blocking of creativity. In science a similar position arises with the notion of absolute truth. In both cases, the attempt to claim an absolute truth about the totality implies an absolute necessity and therefore disposes the mind *never* to yield, no matter what evidence may be found to the contrary. In the face of such an attitude, a genuine dialogue is clearly impossible. The human being is therefore caught up in an unusually rigid infrastructure involving a whole set of assumptions, presuppositions, and practices.

History shows that a true dialogue has never taken place between religions that have appreciably different notions of absolute truth. Indeed, within the same religion, it has seldom been the case that two subgroups have come together again after a doctrinal split. It seems clear that when two such groups differ about the totality, there is no way in which they are able to negotiate their basic differences. At best, they may tolerate each other, but such tolerance is precarious, for sooner or later it may change into a destructive urge to overwhelm the "erroneous" point of view, if necessary by destroying those who hold it. A similar tendency is also present in the secular ideologies that claim to be valid for the totality of life and perhaps even of existence.

But despite these major difficulties, and the implausibility with which many people now view religious assumptions about the totality, it must be recalled that in earlier times religious notions moved entire peoples much more profoundly than science has ever been able to do. These notions entered into great works of art, music, architecture, literature, and poetry. It is therefore important to understand where this extremely powerful impetus came from, and whether or not it may still have a valid place today. This is surely an essential part of the overall challenge to humanity that is being discussed here. To put it in the form of a question: Is the human religious impulse now forever antiquated, or is it something that the human being misses profoundly?

If there is to be a new creative surge it seems clear that it must bring in all three basic dimensions: the individual, the sociocultural, and the cosmic. Indeed with the loss of contact with nature and the general decrease in importance of religion, the civilized world is approaching a state in which it has little sense of direct contact with the totality. It is therefore in danger of losing contact with its key cosmic dimension, just at a time when it needs it more than ever because both the individual and society are overwhelmed with destructive "misinformation." The following sections will therefore explore the question of whether the religious impulse can be freed from its dangerous tendency to become rigidly attached to particular views as to the nature of the totality. For this appears to be the main reason why religions have so often become involved in particularly destructive forms of fragmentation.

The Responses of East and West to the Conditioning of Consciousness

The previous sections have focused mainly on the response of the West to the challenge of "misinformation" within human consciousness. The East, however, over several thousand years, developed its own response, which is only now beginning to penetrate Western culture to any significant extent. In general, it places major emphasis on self-awareness, through inward observation and meditation, in dealing with the conditioning of individual consciousness.

The Western tendency to emphasize outward action and dynamism is thus played down, and in some cases simply rejected altogether, in favor of a tendency toward the suspension of such action, and the encouragement of inward observation and contemplation. This latter is generally regarded as helping to lead to the union of individual consciousness with some kind of cosmic order or ground, as a way of coming to the solution of the problem of human existence.

Such an approach can be seen, for example, in the ancient Chinese notion of the Tao. The Tao emphasizes an inward harmony with nature and with the totality, and implies the minimum of directed purposeful action. (The latter is seen as generally being an interference with the natural order, which is regarded as basically good.) Thus, Lao-tzu, the author of the Tao, praises detached inaction, by saying that "the sage keeps to the deed that consists in taking no action. . . . Do that which consists in taking no action and order will prevail."[5] This theme, that a certain kind of inaction is itself an action, and indeed, the very highest form of action, is one that frequently recurs throughout Eastern culture.

A similar tendency is to be found in Indian culture, starting from ancient times, for example, with the development of Yoga.[6] The form of Yoga most commonly known in the West is Hatha Yoga, which emphasizes bringing about harmony within

5. Lao-tzu, *Tao Te Ching*, Penguin, Harmondsworth, England, 1963.

6. See, for example, *Aphorisms of Yoga*, Bhagwan Shree Pantanjali, translated by Shree Purahit Swami, Faber and Faber, London, 1938.

the body. Although it focuses on movements of the body and therefore does involve a certain kind of outward action, its principal aim is to use such action as an aid to inward perception. To this end, it utilizes a variety of fixed positions (or Asanas), whose main purpose is to make a person aware of the tensions and blockages in muscular responses, which are usually "unconscious." It does this by preventing the habitual movements of the muscles that surround the blockage. These ordinarily serve to hide the blockage by "easing" the tension that would otherwise indicate its presence. Through careful attention to all that is happening, especially inwardly, the nervous impulses that are responsible for the rigid pattern of muscular excitation begin to die away. In effect, a relatively superficial kind of movement in the explicate order is being stopped, and this allows the operation of a much subtler and deeper kind of inward movement in the implicate and generative orders. It is this movement that removes the "misinformation" and heals body and mind.

Such an approach is used in other forms of Yoga, which are concerned with the emotions, the will, the intellect, and so on, as well as also recurring in many other systems of philosophy, meditation, and practice. Thus, in Buddhism, each person is directed, through reflection and meditation, to be aware, moment to moment, of the whole train of his or her thoughts. It is stated that in this process the fundamental "groundlessness" of the self can be seen. In this way a key piece of "misinformation" can be cleared up, i.e., the almost universal assumption that the self is the very ground of our being. This leads ultimately to Nirvana, in which there is a blissful unification with the totality.

A particular form of Buddhism is Zen, which contains the practice of "sitting" in a given position for indefinitely long periods, which is somewhat reminiscent of the fixed positions of Hatha Yoga. If this position, which may become somewhat painful, is maintained, then it is found that all sorts of previously repressed thoughts and feelings begin to come out. It seems that even simple bodily movements that "ease" a sense of tension are deeply involved in avoiding awareness of the unpleasant aspects of the general content of consciousness. As happens with Yoga, the suspension of such outward movements makes possible the deeper inward movement which acts to "dissolve" those features of the rigid infrastructure that are basically responsible for the mental and physical tensions.

Clearly, approaches of this general kind are concerned primarily with suspension of outward activities and impulses in various areas, in order to prevent the mind from "escaping" awareness of the fact that it is conditioned so as to produce blockages of various kinds. The aim is, by careful attention, especially to the inward responses of mind and body to these suspensions of outward activity, to bring the blockages into awareness. This action, which from the outward point of view may be called a kind of "inaction," then makes it possible to clear up the accumulated misinformation of false conditioning that is behind the blockages. Approaches of this kind move in the direction of the transcendental, in that they ultimately merge with the religious-philosophical goal of union of the individual with the ultimate totality. The major emphasis therefore tends toward that which is timeless and beyond the measure of the human mind.[7] In all this, the sociocultural dimension tends to fall into the background and the main interest is in the relationship between the cosmic and the individual dimensions.

A particularly interesting and unusually thoroughgoing example of this approach is from a modern thinker, Jiddu Krishnamurti. His writings go extensively and deeply into the question of how, through awareness and attention to the overall movement of thought, the mind comes to a state of silence and emptiness, without any sense of division between the observer and the observed. In this state, the mind's perception is clear and undistorted. Krishnamurti feels that such a mind is necessary to dissolve the kind of problems that have been discussed

7. Thus, for example, the word *Yoga* has the same root as the English word *yoke* and signifies the joining of the ordinary life to the transcendental ground. See Swami Venkalesanda, *The Concise Yoga Vasistha*, State University of New York Press, Albany, 1984, for a good account of the basic religious-philosophical worldview behind this approach.

throughout this book.[8] However, for him, this is all of secondary importance. His main point is that such a mind is in a suitable state for entering into what may be called the ground of all being and that this is the ultimate meaning of existence.[9]

What is particularly significant in this regard is Krishnamurti's insistence that it is precisely the nonmovement, or inaction, of thought that is necessary for the very being of this other state, which transcends time, space, and anything that can be grasped by thought. Clearly, the principle of the suspension of explicate activity is essentially the same in Tao, Yoga, and Buddhism. It should also be apparent that an immense extension of this principle has also taken place, in the sense that the entire movement of the whole of consciousness is, as it were, suspended to allow the mind to enter the ultimate ground.

It can be seen from this that the East is inclined toward suspension of overt or explicate activity in favor of a kind of movement at subtler levels. A vivid example of this can be seen in the many statues of the Buddha, which suggest perfect repose, balance, and harmony that are not outwardly imposed but which arise from an inward freedom from attachment to anything. In the West, however, what now tends to be emphasized is outward movement, dynamism, and unending transformation. This can be seen in much of its art, such as Michelangelo's statue of David and his painting of the Creation on the ceiling of the Sistine Chapel.

Perhaps both cultures were fairly similar in their early days, but because of what was probably only a small initial difference in tendency, they developed differently, until by now they are far apart. Each of these cultures has had its own characteristic accomplishments, some of which are very impressive. Nevertheless, despite its many creative accomplishments, the Occidental culture is now basically in a state of decline and it does not appear to know of any fundamental way of dealing

8. For this phase, see *The Ending of Time, Thirteen Dialogues Between J. Krishnamurti and D. Bohm*, Gollancz, London, 1984.

9. For this phase, see, for example, J. Krishnamurti, *Freedom from the Known*, Gollancz, London, 1969.

with the problems that brought this about. These problems include not only the dangers of nuclear annihilation but also the destruction of the earth's environment through deforestation, denudement of arable land, pollution, and much more. However, the Oriental culture, which experienced its major creative surge a very long time ago, has also brought about a corresponding range of fundamental problems which it does not appear to be able to solve. While interest in the religious-philosophical line of inquiry has been sustained, at least in India, a great many other aspects of the society at large have remained relatively stagnant over the centuries or have declined. It is true that there has been a recent renewal of energy in such countries as China, India, and Japan. Nevertheless the primary motive force has not come out of the Eastern culture itself but through the adoption of Western science and technology, along with the attitudes and general culture that go with them.

What is clearly needed in East and West is a creative surge of a new order. Such a surge will not be possible while humanity goes on with its current fragmentation, represented by the extremes of Eastern and Western cultures. Nor is it sufficient for each culture to adapt to its own needs certain features from the other that it may find convenient or attractive. For to do this is to still go on with the rigidity of basic assumptions that are characteristic of both cultures. That would lead only to false play and the blockage of creativity. A genuine dialogue between the two cultures is clearly called for, in which there is no holding to fixed points of view, so that a new free and fluid common mind could perhaps arise. Such a mind would have rich new possibilities for creativity, by moving in a different order in the area "between" the extremes of current Eastern and Western cultures. Perhaps ultimately the dialogue could be extended to include the relatively wealthy North and the relatively impoverished South, as well as East and West, so that a truly planetary culture with a socially shared meaning could come into being.

Only the free movement of the mind that can arise in a dialogue will be able to make a major impact on the cultural rigidities that ultimately give rise to the general problems faced by societies everywhere. In this way, for example, East and West could move forward toward a broad "middle ground," between Western dynamism and Eastern suspension of outward activity, as well as between the timeless and the temporal orders, the individual and the social orders, with the cosmic order on one side and the social and individual orders on the other.[10] This would open up a rich new field for creativity in which all could share.

CREATIVITY IN SCIENCE, ART, AND RELIGION

It has been argued that the full unfoldment of creativity requires the ending of rigidity, and therefore of fragmentation in the overall planetary culture. To clarify what this would mean, it is convenient to sum up the cultural life of humanity in three dispositions, approaches, or attitudes of mind: namely, the scientific, the artistic, and the religious.

Although *science* literally means "knowledge," the scientific attitude is concerned much more with rational perception through the mind and with testing such perceptions against actual fact, in the form of experiments and observations. In making such tests, what is crucial is the attitude of acknowledging an actual fact, with properly made inferences, without being caught up in the tendency of the human mind to play false. In most of life this principle does not play a large part. For example, in international relations the actual fact is generally distorted according to what is regarded as most useful or desirable to the state, a position which seems to have first been openly advocated by Machiavelli.

In the light of these implications of the scientific attitude, it seems particularly strange that, in Western culture at least, it has been thought to be necessary only in limited fields. It is as if someone were to say, "In my laboratory I try seriously to acknowledge the actual fact, but in other areas of life, such as

10. It is interesting that Buddhism also seeks a "middle way," though on a basis quite different from that suggested in this book, which is dialogue and the consideration of creative orders that lie "between" the extremes.

human relationships and politics, it is best to play false whenever it is convenient, and to fit the fact to whatever is needed." Clearly it would produce a tremendous revolution if the scientific attitude were genuinely and seriously acknowledged to be valid and necessary for the whole of life. In such a case the very core of the contribution of science to the creative surge would take the form of an extension of the scientific attitude into all human relationships.

Art is based on the Latin root meaning "to fit" and shows up in such words as *article, artisan,* and *artifact.* The history of this word clearly indicates that, in earlier times, there was no separation between art and the rest of life. Thus, an artifact is something made to fit in both an aesthetic and a practical sense. Today, however, a work of art is generally judged as "fitting" in the aesthetic sense alone and this indicates the current state of fragmentation between art and other areas of life.

Art, which includes music, drama, literature, poetry, dancing, and the visual arts, is strongly concerned with beauty, harmony, and vitality. However, more fundamentally, one of its essential meanings seems to be that the "fitting" or "nonfitting" is seen, from moment to moment, in an act of fresh creative perception, rather than through mechanically applied rules as to "what is fitting and proper." In this sense, everything may be thought of as being a kind of art. Thus, in science, the question as to the meaning of a given set of facts and equations has finally to be answered through such a perception, which is basically artistic in nature. And more generally, while much of life is determined by mechanical rules and formulae, it is possible to speak of an "art of living" in which the artistic attitude is conducive to a sustained creative perception.

The artistic attitude is particularly important, with regard to its emphasis on the role of the imagination. Literally *imagination* means "the ability to make mental images," which imitate the forms of real things. However, the powers of imagination actually go far beyond this, to include the creative inception of new forms, hitherto unknown. These are experienced not only as visual images but also through all sorts of feelings, tactile sensations, and kinesthetic sensations, and in other ways that defy description. The ability of Mozart and Bach to sense whole musical works all at once could be regarded as a kind of musical imagination. The activity of the imagination does not therefore resemble a static picture but rather it is closer to a kind of "play" that includes a subtle orchestration of feelings, as well as a sense of intention and will. Imagination is thus the beginning of the entry of creative perception into the domain of the manifest. Moreover, since form is defined by proportion and ratio, the imagination must also possess these in some implicit or enfolded sense. Intuitive or perceptive reason is the act, then, of making explicit the ratio or proportion that is already implicit in creative imagination. In this way reason unfolds from imagination, as indeed some of the leading scientists and mathematicians have implied. Einstein, in fact, has described his experiences of concepts which originate in vague undescribable "feelings" and sensations.

As the imagination "crystallizes," its forms become relatively fixed and give rise to fancy or fantasy.[11] This is the power to form mental images of known kinds, to combine them, and to relate them. Evidently this power is both necessary and useful in, for example, making plans and designs. It corresponds roughly to the power of reason to form concepts of known kinds and then to combine and relate them. But with imagination and reason, the danger is, however, that the "crystallized forms" become excessively rigid when they are regarded as absolutely necessary for the well-being of the psyche and of society. The result is that the mind is caught up in playing false, as it tries to defend them. In this process fantasies are confused with reality, and the formal logical relationships of concepts are confused with truth.

It is clear that a proper appreciation of the artistic attitude should not be left solely to those who specialize in art. An artistic attitude is needed by all, in every phase of life.

---

11. The distinction between creative imagination, called primary imagination, and fancy or fantasy was first made by the poet Coleridge. For a discussion of this point, see O. Barfield, *What Coleridge Thought*, Wesleyan University Press, Middletown, Conn., 1971.

Indeed, harmony can be achieved only through a constantly fresh artistic perception of what is "fitting and proper."

Finally, religion must be examined, which is concerned primarily with wholeness, as can be seen in the word *holy*, which means "whole." Religion tends to emphasize the eternal and timeless, as well as contact with the ultimate ground of being. The word *worship* is based on the same root as *worthy*, and thus implies that act of "giving a very high value," that is, to the source of all that is. Indeed, religion is very much concerned with values, and most religions have argued, either explicitly or implicitly, that it is not possible to give the right values to things unless there is a correct relationship with what has supreme value, that is, with God. The tendency of human beings to put a supreme value on *something* is indeed so great that those who give up religion generally tend to attribute this sort of value to something else, such as the welfare of the state, or the happiness of the individual. It seems clear, therefore, that either humanity goes on with religion or it decides to drop it, in which case it still must deal with the question of what, if anything, has supreme value.

As with science and art, a truly religious attitude has to be free of rigid commitments in the tacit infrastructure of consciousness, so that all that is done comes out of creative perception. Religion could then be deeply concerned with an inquiry into whether, and how, a human being can come into contact with the ground of all. But at the same time, it would also be deeply concerned with the sociocultural and individual dimensions of human life, and especially with the question of how and why human beings have such a strong tendency to be caught up in playing false, which, in current religious terminology, is called "sin" or "evil." Above all, a religious attitude has to be compassionate, to acknowledge the ultimate value of the human being, and to realize that each individual shares in the general human conditioning over the ages to confuse and to play false.

Throughout history, however, it must be admitted that religions have tended to be caught up in all kinds of self-deceptions and in the exploitation of others. The origin of this can often be traced to the uneasy balance between religion and secular powers, a tension which is now giving rise to similar problems in science. But more fundamentally, the role of absolute belief, whether in some notion as to the nature of God or to the ultimate nature of being, has given rise to the most serious problems. Any belief in some definable form of the absolute entails the notion of absolute necessity, which leads to an unyielding attitude in which basic assumptions are nonnegotiable. In the grip of this attitude, people find themselves compelled to fight to the death over their different views, and even religions that proclaim love for all have become involved in the spread of hatred.

In this context it is useful to investigate more carefully what is meant by belief. The word itself is based on the Teutonic Aryan word *lief*, which means "love," so that what is believed is "beloved." The danger in belief should therefore be clear, for when the "love" for a set of assumptions and their implications is strong, it may lead to playing false in order to defend them. The end result is inevitably destructive. On the other hand, it should be recalled that belief also implies trust, confidence, and faith in the essential honesty and integrity of something—for example, a person, an institution, a cultural activity, and ultimately life and creativity. Without such belief, the serious and sustained commitment that is necessary for creativity will not be possible.

What is needed is clearly a middle ground between the extremes of credulous belief, aimed at making people feel happier and more secure irrespective of whether the beliefs happen to be correct, and a total skepticism which results in a cynical attitude to everything. It may be possible, for example, to entertain a range of assumptions with trust and confidence, in which none is so sacrosanct as to lie beyond serious questioning. If such an approach were an integral part of the religious attitude, then the basic conflict between religious and scientific attitudes would cease. Indeed a religious inquiry would be just as open as a proper scientific inquiry.

With such an attitude the principal objections to religion that were discussed earlier would vanish. For the religious approach would no longer imply any absolute commitment to rigidly fixed notions as to the nature of the totality. Indeed such an attitude could be expressed by the statement, Whatever we say the totality is, it isn't—it is also more than we say and different from what we say.

Originally, science, art, and religion were not distinct but were inseparably united. Considering that these three have such a deep significance throughout human history, it seems clear that the present gulf between them must have a harmful effect in the generative order of the consciousness of humanity. But there is no intrinsic reason why these three attitudes have to be separated. Rather, while one of them may be emphasized in a particular activity, the others must always be present although, for the moment, they may be in the background. Nevertheless humanity has become conditioned to accept such a rigid separation. What is clearly needed is a dialogue between these attitudes, in which sooner or later they can all come into the "middle ground" between them, which will make available a new order of operation of the mind with rich possibilities for creativity. The opening of such a dialogue could play a crucial role in freeing the consciousness of humanity from one of the most significant blocks to creativity in its tacit infrastructure.

A New Order of Creativity

It should now be clear that the creative surge that was called for at the start of this book cannot take place in science alone. Rather, every phase of human life has to be involved. Something along these lines must have taken place during the Renaissance in a radical transformation that included science, art, and a new view of humanity, culture, and society. What is needed today is a new surge that is similar to the energy generated during the Renaissance but even deeper and more extensive.

It is not appropriate here to give a detailed program for how such creativity could be brought about. In fact the very nature of creativity precludes any such program, which would have to include within it a tacit definition of what creativity is, or at least some assumption of how it is to be achieved. Such definitions and assumptions would in effect become rigid commitments in the tacit infrastructure of consciousness, and would sooner or later constitute blocks to the very creativity that they were designed to elicit.

Since the potential for creativity is natural, the principal question is that of revealing the rigid assumptions in the tacit infrastructure that block this creativity and then being able to dissolve them. This will bring about a transformation in the order of awareness and attention, which enables the mind to respond freely to fresh creative perceptions. A number of ways have already been suggested in which this question can be explored through dialogue, as well as Eastern approaches that are aimed at bringing about a degree of self-awareness. What is essential, however, is for the mind to move into the broad "middle ground" between extremes. Indeed creative intelligence may quite generally be regarded as the ability to perceive new categories and new orders "between" the older ones, which are, in this case, the disjointed extremes. In doing this, it is not just "mixing" the extremes, or "selecting useful bits" from them. Nor is it a matter of engaging in some "middling" or "mediocre" type of action. Rather, as in the case of regular orders of low degree and chaotic orders of infinite degree, what lies between is a new domain for creativity, which is qualitatively different from either of the extremes.

It is important to extend creativity beyond the spheres in which it is traditionally supposed to lie. This should include not only the exploration of overall global sorts of domains, such as science, art, and religion, but also the more limited activities of everyday life. For example, if there are serious problems in human relationships, it is necessary to become aware of rigid assumptions in the tacit infrastructure of consciousness that are giving rise to them, and to cease to be caught up in these assumptions, along with the "emotional charge" that goes with them. In this connection, recall how, in her work with Helen Keller, Anne Sullivan had to become aware of tacit assumptions concerning the roles of language and concepts that were taken for granted. Her mind was freed to respond creatively in new ways, in which previously "insoluble" problems were "dissolved."

It is clear that something like a pervasive creativity should be present in all aspects of life. To bring this about it is necessary to question, very seriously, the current assumption that creativity is needed only from time to time, and then only in certain special areas, such as art and science.

A very important question is that of how this new order of creativity can ever get started. For both individually and socially, consciousness is rigidly conditioned by a host of assumptions that lead to their own concealment through false play. In the resulting confusion and illusion, the mind is not even able to be aware of these assumptions, or to give proper attention to them. Various ways have already been suggested in which the mind may be able to "loosen" some of these assumptions. The essential point, however, is that *any* kind of free movement of the mind creates the opportunity for revealing and loosening the rigid assumptions that block creativity. Here it should be noted that this blockage is never *total*, for everyone has some areas that are still open to free and honest inquiry, in spite of the effects of a lifetime of conditioning in a society that generally discourages creativity. It is therefore important to discover where these areas are. Wherever a person finds that he or she can be creative, this will be a good starting point.

Usually, when a person finds the possibility of creativity in certain areas, he or she is happy to go on with the resultant activity. But the crucial point is not just to stay with this particular activity but to give attention to the creative movement itself. For as in the case discussed in Chapter 5 of observing the stream, it is possible for a similar movement to go, as it were by analogy, to other areas. What is especially significant is that whatever its content, this creative movement has the kind of passionate intensity and vibrant tension that is able to bypass and even to dissolve the blocks to creativity. In this way, a far-reaching and penetrating movement can start, which can eventually have a profound effect in all areas of life. Moreover any person who reveals a sustained creativity throughout his or her life will tend, also by a kind of analogy, to bring about a similar movement in other people.

The key point is that it is not enough to be interested only in the particular results or products of creativity, as they are manifested in limited fields. The general decline of creativity in a society is itself a kind of "illness" which must ultimately bring about its destruction. It is therefore crucial that whatever creativity remains shall be turned toward the destructive "misinformation" that is blocking and gradually choking off the natural potential for creativity.

SUMMARY AND OUTLOOK

At the beginning of this book a new creative surge was called for in order to meet the extraordinary challenge that is now facing the human race, which has implications in almost every field of activity. The book began with an investigation of the nature of creativity and what impedes its operation by considering creativity in science. In these examples it was clear that the essence of the creative act is a state of high energy making possible a fresh perception, generally through the mind. This is blocked by the rigidly fixed tacit infrastructure of consciousness, which cannot properly respond with "free play" to such perceptions. Instead the mind "plays false" to create the illusion that no disturbing new perceptions are needed.

The discussion was then extended to show that, in science at least, free play in *communication* is also necessary for full creativity, and that this, too, is blocked by the rigid content of the tacit infrastructure of the general consciousness.

In discussing the whole question of order, which plays a key role in creativity, it was shown that between two extremes, of simple regular orders and chaos, there is a rich new field for creativity. What is particularly significant here are the generative and implicate orders. For through these, it becomes possible to understand the unfoldment of creativity from ever subtler levels, leading to a source that cannot be limited or grasped in any definable form of knowledge or skill. This source cannot be restricted to particular areas, such as science and art, but involves the whole of life. Therefore, the creative surge that is called for will have to be general and pervasive, rather than limited to special fields.

It is crucial in this connection to understand that "errors" or "misinformation" that are enfolded deep in the generative order may have extremely wide-ranging and serious negative consequences. Thus, if there are rigid ideas and assumptions in the tacit infrastructure of consciousness, the net result is not only a restriction on creativity, which operates close to the "source" of the generative order, but also a positive presence of energy that is directed toward general destructiveness. A clearing up of such "misinformation" is therefore needed if this energy is to be freed from its rigid and destructive pattern, so that it may respond properly in order to unfold creative perception in manifest forms. One of the main purposes of this book has been to draw attention to the key importance of liberating creativity, if human life is to have a worthwhile kind of survival. Indeed, of all the discussions that have taken place over the various crises that face humanity, this essential factor has not been adequately emphasized.

A number of possible ways of clearing up the "misinformation" were described. In the West, various forms of psychology are among the principal approaches to treating the human being of such misinformation in the dimension of the individual. In both the East and the West, various philosophical, religious, and mystical approaches have also been developed for clearing up the cosmic dimension in a similar way. However, with all these approaches, an essential factor has been given too little attention, namely, the sociocultural dimension. A major part of the mind's misinformation arises in this area, and cannot be properly cleared up through the individual and cosmic approaches. Moreover since a large part of our very being lies in this dimension, confusion in this area can have particularly disastrous consequences. Vice versa, to clear up this area can liberate the considerable energies that are associated with a properly operating consensual mind. Such a mind is moved by a spirit of impersonal friendship and is open to creative intelligence in ways that are beyond those accessible to the individual. This is why dialogue has been so strongly emphasized in this book, since it can "loosen" the collective, sociocultural rigidity that holds all of us in its grip.

As long as this general sociocultural rigidity prevails, communication on fundamental issues will be blocked, in the sense that people will not be able to listen to each other seriously whenever basic questions are raised. The result is a proliferation of fragmentation, which so pervades society today and which appears to have been characteristic of most if not all known societies. Without an approach that addresses itself directly to "misinformation" in the sociocultural dimension, it seems unlikely that any of the approaches that operate through the other basic dimensions of humanity can get very far.

Consider, for example, a hypothetical individual whose consciousness had been "cleared up" both in the individual and the cosmic dimensions. Although this person might be a model of wisdom and compassion, his or her value in the general context would be limited. For because of "unconscious" rigidity in the general infrastructure, the rest of humanity could not properly listen to this person and he or she would either be rejected or worshiped as godlike. In either case there would be no true dialogue at the social level and very little effect on the vast majority of humanity. What would be needed in such a case would be for all concerned to set aside assumptions of godlike perfection, which makes genuine dialogue impossible. In any case, the truly wise individual is one who understands that there may be something important to be learned from any other human being. Such an attitude would make true dialogue possible, in which all participants are in the creative "middle ground" between the extremes of "perfection" and "imperfection." In this ground, a fundamental transformation could take place which goes beyond either of the limited extremes and includes the sociocultural dimension.

It is particularly important to emphasize dialogue and the sociocultural dimension here, because they have generally received so little attention in this context. But such emphasis should not result in the neglect of the other dimensions.

Indeed, in any creative activity, all three dimensions will have to be present, though at any given moment, one may be accentuated in relationship to the others.

It is essential that in the long run, the context of dialogue should include not only discussions relevant to the sociocultural level but also discussions of the life and problems of the individual, and of the cosmic context, both as this latter is revealed scientifically and as it is felt psychologically and religiously. Vice versa, creative activity in the other two dimensions will have to have in it the spirit of the dialogue, in which many points of view can be held in suspension, and in which the creation of a common meaning is a fundamental aim.

In all three of these basic dimensions of culture, the essential need is for a "loosening" of rigidly held intellectual content in the tacit infrastructure of consciousness, along with a "melting" of the "hardness of the heart" on the side of feeling. The "melting" on the emotional side could perhaps be called the beginning of genuine love, while the "loosening" of thought is the beginning of awakening of creative intelligence. The two necessarily go together. Thus, to be "warmhearted" and "generous" while keeping ideas rigid will lead only to frustration in the long run, as will intellectual clarity that is allied with cold hard-heartedness.

The ultimate aim of this book has been to arouse an interest in the importance of creativity. Whoever sees this importance will have the energy to begin to do something about fostering it, in ways that are appropriate to the special talents, abilities, and endowments of that person. All great changes have begun to manifest themselves in a few people at first, but these were only the "seeds" as it were of something much greater to come. We hope that this book will not only draw attention to all the questions that have been discussed in it, but will actually begin the liberation of creative energy in as many of its readers as possible.

INDEX

Printed in the United States
by Baker & Taylor Publisher Services